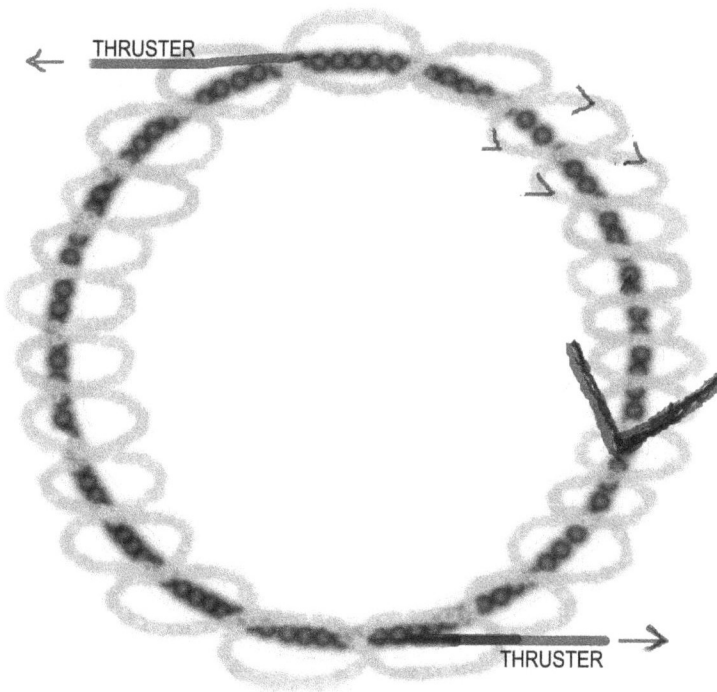

Schematic diagram of a braided ion accelerator ring and a quark plasma ring. These are the major components of a quark drive engine for a starship. A starship has two such engines. (Arrows indicate the direction of flow in the rings.)

The large thick ring holds accelerating quark plasma. The ring that oscillates (braids) "around" it has accelerated heavy ions such as U^{238} ions. The ion ring self-collides at points along the quark plasma ring to create and feed quark-gluon plasma to the quark plasma ring where the quark plasma is in turn accelerated. The two thruster ports siphon quark plasma from a quark ring generating a type of "ion" thrust. Either one or the other port siphons the quark ring depending on whether we want "forward" thrust or "rearward" thrust.

Effectively a quark drive is like an ion drive except that the speed and complex momenta of the ejected quark plasma stream enables faster-than-light motion for a starship.

There is another similar quark engine (with opposite flow in its rings) on the starship that together with the above engine provides forward or rearward thrust from diagonally opposite tangent points on the two engines. See appendix B for details.

TO FAR STARS AND GALAXIES

SECOND EDITION OF
BRIGHT STARS,
BRIGHT UNIVERSE

Stephen Blaha, Ph.D.

Pingree-Hill Publishing

ISBN: 978-0-9819049-5-5

Cover Credits
Cover by Stephen Blaha © 2009. The cover includes an illustration of a "Barred Spiral Milky Way Galaxy" created by R. Hurt (SSC) of JPL-Caltech for the NASA GLIMPSE Survey Team. Used with permission of NASA.

rev. 00/00/02

To John, Natasha, and Stephen

Some Other Books by Stephen Blaha

The Algebra of Thought & Reality: The Mathematical Basis for Plato's Theory of Ideas, and Reality Extended to Include A Priori Observers and Space-Time Second Edition (ISBN: 9780981904931, Pingree-Hill Publishing, Auburn, NH, 2009)

A Complete Derivation of the Form of the Standard Model With a New Method to Generate Particle Masses SECOND EDITION (ISBN: 9780981904900, Pingree-Hill Publishing, Auburn, NH, 2008)

Physics Beyond the Light Barrier: The Source of Parity Violation, Tachyons, and A Derivation of Standard Model Features (ISBN: 0974695874, Pingree-Hill Publishing, Auburn, NH, 2007)

Quantum Theory of the Third Kind: A New Type of Divergence-free Quantum Field Theory Supporting a Unified Standard Model of Elementary Particles and Quantum Gravity based on a New Method in the Calculus of Variations (ISBN: 0974695831, Pingree-Hill Publishing, Auburn, NH, 2005)

Quantum Big Bang Cosmology: Complex Space-time General Relativity, Quantum Coordinates™ Dodecahedral Universe, Inflation, and New Spin 0, ½, 1 & 2 Tachyons & Imagyons™ (ISBN: 0974695815, Pingree-Hill Publishing, Auburn, NH, 2004)

The Metatheory of Physics Theories, and the Theory of Everything as a Quantum Computer Language (ISBN: 097469584X, Pingree-Hill Publishing, Auburn, NH, 2005)

A Unified Quantitative Theory Of Civilizations and Societies: 9600 BC - 2100 AD (ISBN: 0974685858, Pingree-Hill Publishing, Auburn, NH, 2006)

Available on bn.com, Amazon.com, and other web sites as well as at better bookstores (through Ingram Distributors).

Preface to Second Edition

This book is the second edition of the ground-breaking book on faster-than-light travel *Bright Stars, Bright Universe: Advancing Civilization by Colonization of the Solar System and the Stars using a Fast Quark Drive* which described a totally new approach to spacecraft propulsion that will enable Mankind to travel to the stars, and beyond, to the galaxies of the known universe. In this edition we add Appendices to the original edition (together with a few minor changes and corrections to the First Edition) on Seeing and Navigating through the Cosmos on superluminal starships; on Complex Thrust using new "Braided" Accelerators; on Suspended Animation to keep the biological clocks of starship occupants in sync with earth time; on Engineering very long life starship and life support components; on Robot guidance and robot exploratory starships; and on Fuel consumption on starships.

At the time of this writing the United States appears headed towards a deficit of ten or eleven trillion dollars over the next ten years. It appears that a likely cost for starship R&D would be of the order of half a trillion dollars spread over perhaps thirty years – a small amount relative to the projected US deficit – with an enormous reward for success. On a yearly basis this expense averages to about seventeen billion dollars per year. Since the NASA budget is eighteen billion dollars a year, the cost of the starship program is comparable and quite affordable. The formation of an international consortium would further reduce the per country costs. If the United States paid sixty per cent of the costs then its share would be about ten billion dollars per year. This venture is of great significance to the future of Mankind. The cost is relatively small; the benefits are potentially enormous!

PREFACE

In a series of recent books we explored the theoretical importance of superluminal (faster-than-light) particles called tachyons in the derivation of the form of the Standard Model of Elementary Particles. We also showed some of the remarkable features of superluminal particles such as reverse fission, length dilation, and time contraction – quite the opposite of sublight phenomena.

In this book we examine the possibility of superluminal starship propulsion for interstellar exploration and colonization. We begin the book with a statement of the earth's present condition and the need to expand into space or face a slow decline into a cultural and social miasma. Then we develop a new plan for the exploration and colonization of this solar system since it appears that current announced plans have missed the mark and do not have an overall long-term approach.

Since this solar system lacks enough planets that are congenial for human life we consider the possibility of colonizing planets around other stars. Many plans have been proposed for interstellar spaceships – starships. But they are usually impractical for some good reason(s).

In this book we will consider superluminal (beyond light speed) starships based on quark tachyon dynamics. They provide a practical means of starship propulsion although much R&D must be done before the proposed starships can be built. Once built they will support speeds up to 5,000 to 30,000 times the speed of light *and far beyond* making the universe Mankind's backyard. Travel times to other galaxies could be as short as a few months in earth time. *Superluminal starships open the door to a vast expansion of humanity into the universe in the large!*

As world population grows, and human social needs also grow, the capacity of the world to mount a major space effort will diminish. So a major space initiative with a well-thought out game plan is required *now*, while we still have the resources.

CONTENTS

iv

FIGURES

1. Earth: Now a Zero Sum Planet

1.1 21st Century Earth

The year 2008 was a remarkable year – not because of the armed conflicts that are taking place in various parts of the world – but because we see for the first time that the earth is a "zero sum game."[1] The gains experienced by one country were counterbalanced by the losses of other countries. In the past decade the United States has "lost" a substantial amount of its capital through foreign trade to Asian and European countries. The prosperity of these countries has increased. Now the United States is facing massive economic problems because of the loss of many industries and the outflow of its wealth. A consumer country without a strong manufacturing base becomes an economic bleeding corpse.

The growth of prosperity in Asia and Europe is a welcome event but it has become clear that much of it was at the expense of the United States and other wealthy countries. Most people hope that all the countries in the world will eventually reach a common high standard of living not unlike that which the United States has enjoyed in recent decades on average. Such a world would have much less reason for warfare and a world at peace would be a likely prospect.

However this dream cannot be accomplished for several reasons. Given the world's population of six billion and the fact that the United States consumes 40% of the world's production while having only about 5% of the world's population, world production would have to increase by a factor of roughly 8 for the average world standard of living to be

---

[1] A "zero sum game" is a game in which one person's gains are counterbalanced with another person's losses. Thus the sum of all players "chips" remains constant and the change in the total sum as the game progresses is zero. The commodities markets (in gold, grains, pork bellies, currencies and so on) are considered to be zero sum games. The gains (profits) of one trader are the losses of another trader(s). Thus the total sum of all traders' investments in a commodities market is unchanging.

that of the US. Eight times more food, eight times more housing, eight times more industry, and so on. Clearly, an impossible task for the earth to achieve and sustain.

Now we must confront the fact that the world today is at its environmental limits. The degradation of the environment at today's production levels is obvious: water and land are increasing polluted (poisoned). In particular, the oceans are being polluted with heavy metals, plastics, sludge and so on. We are entering a period of man-made global warming, increasing air pollution, increasing birth defects and educational disabilities, and other signs of an overstressed and precipitously declining environment. As the population grows to nine billion in 2050 we can expect to see conditions worsen.

And the terrible fact is that we cannot change these trends without massive efforts that will cause a major lowering of the standard of living for all nations. *These facts "everybody" feels but political leaders in every country cannot voice them without causing major population unrest.* Can the leaders of China or India say they can never reach the American standard of living without causing their peoples to lose hope and perhaps turn to unscrupulous demagogues who promise anything to achieve power? Can an American president say that universal lifetime health care, while maintaining current high standards of medical quality, is not economically feasible over the long term?

Clearly world leaders avoid stating the realities of the world situation and put in place "programs of hope" that promise eventual benefits but in reality cannot lead to the nirvanas they boldly promise.

The most significant and most understated problem facing the earth is the need to balance production with effective waste disposal. Manufacturing in all its facets is producing products, which often require waste disposal methods whose costs are comparable to the costs of their production.

1.2 Zero Sum World Economy

Until recently growth and development were the engines of prosperity. Now we must change to a zero sum economy where *"every dollar spent on production must be matched by a dollar spent on recycling and environmental protection."*

Further the Chinese program of population limitation should be taken up by the other nations of the world in an appropriate form. The western nations of Europe and North America have a stable or declining population if the immigrant population is not counted. The countries in Asia, South America, Central America, and Africa need to strongly encourage population limitation with the spirit of the Chinese model but not necessarily with the methods of the Chinese government.

An eventual downsizing of the earth's population to about one billion appears to be required if we wish a high standard of living for all combined with a clean, stable world environment.

1.3 The Effect of a Declining Population

Downsizing the population of a nation or a planet is an undertaking fraught with dangers. Japan is experiencing some of those difficulties as its population declines, and ages. Economic problems include the cost of supporting an aging population. And the major burden falls on the working, youthful part of the population whose percentage of the total population is decreasing.

There is a more subtle cultural issue associated with aging. The creativity and cultural progress generated by the more youthful part of the population is less strong as the relative proportion of younger people declines.

In the case of our world, the population will grow from the current six billion to nine billion in 2050 (a figure which cannot be changed except by drastic events). Then the population must decrease to about one billion perhaps over one hundred and fifty to two hundred years to achieve an eventual, maintainable, high standard of living for all. One can only expect that near totalitarian means will have to be used to achieve this goal. The result of this plan will be to profoundly affect the world's economy and the world's ability to engage in major projects such as space exploration and scientific research.

At present large scale scientific and space projects are carried on by nations with large populations with the economic wherewithal to pay for these projects. Large-scale projects cost little for large countries on a per capita basis. A small population necessarily will have a "small" overall economy and will not be able to finance major projects such as space exploration at a low per capita cost in the manner that we see currently unfolding in the major US, EU, Chinese, and Indian space programs.

Lastly, if the earth's population is reduced to about one billion it is possible that "overshoot" may occur and the population might further decline to unacceptable levels. A historical example of this possibility is the latter phases of the Roman Empire when many areas including Italy experienced large declines in population. Augustus Caesar noting the decline in birth rates amongst the Roman population started the practice of paying bonuses to parents for having children. More recently, France, Germany, and other modern European countries have experienced continuing population declines.

Thus it seems best to make major space efforts now when the world can most afford it rather than make minor showcase efforts—an accurate description of current efforts.

1.4 An Alternative: Beehive Earth

If we fail to begin major efforts soon then one can expect the earth to eventually have a petrified civilization that will ultimately lead to a "Beehive-Earth" with a stratified society with little or no hope for the future.

With the hope for a better future destroyed it is possible that the world faces a period of fragmentation, and perhaps warfare, competing for resources and clean water.

1.5 Environmental Protection will Never Completely Work

There is a relatively strong conservation movement in the United States and in other countries around the world. While this movement will achieve victories in the struggle to protect the environment from further deterioration, the shear size of earth's population makes it impossible to prevent environmental deterioration now and for the foreseeable future.

Much of the world is in a growth phase currently, particularly India, China and Southeast Asia. The result of this growth phase is increasing pollution. Eventually an equilibrium point will be reached where growth will give way to environmental maintenance and recovery. The recycling of waste materials will in itself become a major task for the world's population. The consequence can only be a decline in the world's standard of living.

1.6 Hope: Colonize Space

Approximately twelve thousand years ago the major ice ages ended and the scattered parts of Mankind began to develop societies in various parts of the world. After domesticating animals and plants the population density of Mankind, and its wealth, reached a point where civilizations became possible about 3000 - 4000 BC. Interestingly civilizations[2] appeared more or less simultaneously on the continents of Asia, Africa and South America.[3]

THE EXPANSION OF CIVILIZATIONS

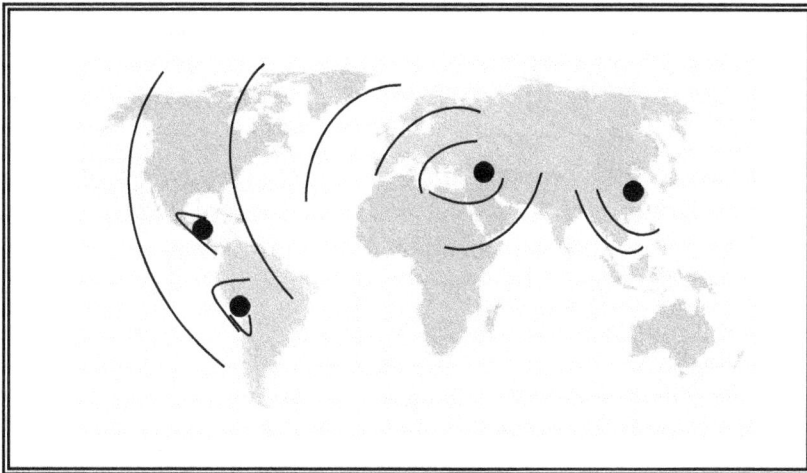

Figure 1.1. Some mushroom rings of civilizations.

Since the first civilizations, Mankind has expanded enormously in population, and in culture and technology.[4] We now have a world

---

[2] Much of the material in chapter 0 and chapter 1 first appeared in Blaha (2006).
[3] In Asia the Yellow River civilization of China, the Mideast civilizations, Egyptian civilization, and the Caral civilization of Peru (recently discovered by Dr. Ruth Shady) all appearing within several hundred years of 3,000 BC. Both the Caral civilization of Peru and Egyptian civilization built great pyramids on a similar scale.
[4] It is this author's opinion that the somewhat astounding advances in Science and Technology—particularly in the past one hundred years—are actually only a beginning and that continued progress in these areas will lead future generations to look upon the present state of these fields of

teeming with people and an explosive growth in communications (satellite communications, the Internet and so on) that seems to be suggest that a world civilization is in the making. Together with that growth we clearly see that Mankind is pushing the limits of the earth's ability to sustain it. The land, air and oceans are undeniably polluted.

Since we have reached the limits of the earth the only viable option is to expand into space to new virgin planets.

The first question that we must address is "Why?" We do not presently appear to have any great practical reason to reach out from the earth.[5] And the growth of the human population, from 6 billion or so individuals to many more billions through the colonization of nearby moons and planets in space, is not in itself appealing.

The answer seems to lie in an old truism, "What does not grow, decays." Thus there appears to be an inherent need for new ground for the continued progress of Mankind. That is the best answer to the "Why?" of space colonization.

Until the 20th century, there was always new ground for the growth of civilizations. As we have seen earlier, civilizations radiated outwards from their original locations: in South America, Anatolia, the Mid-East, India, China, Africa and so on. Those civilizations, and parts of civilizations, which were at the leading edge of growth – culturally and technologically – were on the "new ground" for the most part. We call this the *mushroom ring* effect. Civilizations grow outward in an expanding cultural and geographic ring. We have seen these rings of growth in the evolution of Chinese civilization from Sinic civilization, and the expansion of Hellenic civilization from Greece to Italy, the far shores of the Mediterranean, Western Europe, and eventually the Americas; and also into southwestern Asia and India. Other civilizations such as Islamic civilization also show significant mushroom ring effects.

There are easy explanations of the growth differential between "old ground" and "new ground." New ground often tends to be more

---

endeavor as primitive and barely a start. The exploration of the possibilities of the human mind is also in a very preliminary stage in the author's view.

[5] The other planets, moons and planetoids do not contain any exotic elements or chemicals that are not found on earth or synthesizable on earth. Excepting possibly for Astronomy, the search for extraterrestrial life, and certain solid state physics experiments that require zero gravity, earth-based experiments are as feasible as off-earth experiments and usually substantially less expensive.

fertile, has not had the environmental destruction of old ground,[6] and tends to attract the more daring, innovative segments of the population who interact to produce new solutions for the issues facing humanity.

Today, in the absence of new ground, there appear to be three ways in which Mankind can grow: the use of technology to restore the environment and lower the needs of Mankind for raw materials making the earth more habitable;[7] an expansion into space – a societal challenge that could bring about a major advance in civilization; and an expansion of our use and knowledge of the human mind to achieve a higher level of culture. The third possibility does not seem to be practical for the mass of humanity and would lead to a civilization of physical self-denial amongst growing environmental disaster. Therefore we will not consider this possibility further.

A combination of the first two possibilities seems to provide the optimal answer: improve the earth and go into space.

1.7 Space and Technology as "The New Land"

One possible strategy for achieving a good standard of living for the bulk of the world's population is miniaturization. Technology has made great strides in electronics and computers to make smaller and smaller components and gadgets.

Unfortunately there are certain items: homes, cars, and necessarily human-sized items that cannot be miniaturized. As a result technology can only partially reduce the resource needs of the world's population.

Thus earth-based technology cannot solve the world's problems.

A massive expansion into space is the only "humane" solution for the earth. The requirements for this expansion are the construction of a large fleet of spaceships that can transport large numbers of people –

---

[6] Among the few places where human occupation has apparently improved the environment are certain South Pacific islands were the inhabitants have developed farming practices that have improved the fertility of the soil. Israel has restored the fertility of much of its farmland through innovative farming techniques. Egypt also has restored parts of its northwest (south of Alexandria) using water from the Aswan Dam and innovative farming techniques.

[7] The growing miniaturization in electronics and computer equipment, and possibly nanotechnology, could make it possible for humanity to live well with lesser use of resources. Freeing the massive dependence of industry on water can make more water available for agriculture in desert areas. Genetic engineering could lead to crops that feed more people and require less water and fertilizer. New technology might be able to reverse much of the environmental degradation that has even polluted the (until recently) "boundless" oceans.

particularly to Mars. The technology of suspended animation must be developed so colonists can sleep on their way to Mars. And, firstly, and most importantly, Mars must be upgraded to a human livable planet through bombardment with water asteroids from the asteroid belt and oxygen generation using genetically engineered plant life. Recent studies have shown Mars already possesses large quantities of water, carbon dioxide and iron (for construction). Nuclear reactors to provide power for mining and construction can be widely used on Mars in the absence of any significant population. So the earth has a solution available if it grasps the opportunity.

This solution cannot solve the earth's overpopulation and pollution problems but it can ensure the spread of humanity to nearby moons and planets so that humanity can expand and thrive in new environments. These new environments will stimulate the expansion of human endeavors to higher levels of achievement. Earth is the birthplace and nursery. The universe is the college of Mankind.

1.8 Why the Urgent Need for Space Colonization Now?

Today the nations of the earth have the resources necessary to launch a major space effort. At the time of this writing (Winter, 2008) the slowdown in the world's economy is causing major unemployment and a decline in consumer demand. Major nations such as China and the United States have announced large public works programs to build and rebuild their infrastructure: roads, bridges, dams, and so on.

They should also consider building spaceports, which are few in number, and factories and support facilities for space rockets. These future-looking projects would furnish employment, and create an infrastructure for space travel. Then a series of manned flight efforts to reach Mars and the asteroid belt, and begin the transformation of Mars should be started. An effort of this magnitude in itself would provide a major boost to the world economy and perhaps redirect us from weapons and armaments to constructing a greater future for Mankind.

Should this opportunity be missed, or begun in a slow fashion, it is possible that the needed resources may not be available fifty or one hundred years from now.

The earth is becoming much poorer.

2. Stages of Solar System Space Exploration

Numerous proposals have been prepared for the exploration and colonization of nearby moons and planets as well as for large space stations.[8] While many of these are sensible and feasible it appears that the full utilization of available technology and resources has not been properly considered in many of them. Partly, this is due to "emotional" commitments to certain technologies. Partly this is due to the comfort of using previous successful approaches.

Our proposed plan is based on multiple technologies that are suited for each stage of space flight. Because of the success of the American program to place a man on the moon directly from the earth we tend to ignore the fact that it is generally more sensible to engage in space flight in stages with each stage most efficiently adapted to the needs and conditions of that stage.

The solar system stages as we see them are:

1. Earth to Earth orbit
   Mechanism: Chemical Rockets, Space Guns, and Rail Guns
2. Earth orbit to moon orbit
   Mechanism: Chemical Rockets
3. Moon orbit to/from a moon base
   Mechanism: Nuclear Rockets, and Rail Guns
4. Moon orbit to Mars or moon orbit to the asteroid belt
   Mechanism: Nuclear Rockets
5. Eventually, Moon orbit to Venus, Mercury, the asteroid belt, and the moons of Jupiter and Saturn
   Mechanism: Nuclear Rockets

---

[8] Various approaches are discussed by Zubrin (2000) and Freeman (2009). The author feels the approach proposed here is the most feasible.

The latter part of this book is devoted to travel to the stars. A new technology is needed for star travel in a meaningful way. Most of the proposed methods are at best "token" travel without significant benefits for Mankind. These methods typically require enormous resources beyond the scope of earth's available resources for the foreseeable future.[9] We will propose a new method for star travel based on an extension of Special Relativity to faster-than-light travel. Our approach leads to potentially enormous starship speeds. We will consider examples of speeds of 5,000 and 30,000 times the speed of light. Much higher speeds are also possible. These speeds reduce star travel and even galaxy travel times to days and months – not generations. The Andromeda galaxy is "just around the corner" in our starships.

---

[9] See Matloff (1989) for a detailed discussion of the requirements of various proposals.

3. From Earth to Earth Orbit

The relatively strong gravity of the earth has locked Mankind to the planet until the 1950's. The only viable method to escape the earth that has been used since the 1950's was the use of ever-larger, chemical propellant rockets. Rockets have enabled us to reach as far as the moon with manned space flights and as far as the edge of the solar system with unmanned probes.

From the 1960's through the 1980's other possible solutions were considered including nuclear rockets and "space guns." These approaches were not implemented for a variety of reasons discussed below.

Recently other approaches, and variations on rocket technology, have been considered by private companies as well as by the US, Russia, China, Japan and India.

In this chapter we will outline some of the approaches to escaping the earth's surface to space. The reader is directed to Matloff (1989) and Freeman (2009) for more detailed discussions.

3.1 Chemical Rockets

Chemical rockets and modified "rocket planes" are prohibitively expensive if one wishes to move people and equipment in bulk to space and/or earth orbit. Rockets restrict travel to a few essential personnel, and equipment for satellite communications and scientific experiments. The initial steps in rockets to space were based on solid fuel rockets that uses ammonium perchlorate as a fuel component. Ammonium perchlorate rockets are not only extremely expensive but also are environmentally dangerous.[10] Large liquid fuel rockets are now being

---

[10] Perchlorates are neurotoxins that have been implicated in health problems such as tumor growth on the thyroid gland, mental retardation birth defects, and learning disabilities in children. Perchlorates have been found in trace amounts worldwide. Areas near most of the 12,000 military installations and rocket fuel plants in the US exceed the EPA standard of 24.5 parts per billion in

developed that use hydrogen and oxygen as propellants. Hydrogen and oxygen combine to produce water – environmentally safe. But the production of hydrogen requires large amounts of electricity, which currently is generated by coal-fired plants. So hydrogen fuel in vast quantities has a significant environmental impact. Despite being environmentally safe liquid fuel rockets require so much energy to produce that they cannot provide "mass transit" to space.

3.2 Nuclear Rockets

Nuclear rockets are significantly more economical than chemical rockets. However, the possibility of an accident on a nuclear rocket has led space agencies to conclude the risks of nuclear rockets are too high for use for transit from the earth's surface to space. The Chernobyl disaster shows the impact of a nuclear rocket accident in flight could be a calamity for a large area of the earth's surface.

However, nuclear rockets can play an important role in travel in the solar system so we will consider them in chapter 5.

3.3 Chemical Space "Guns"

Another approach to large-scale travel into space is through the use of "space guns" of the type proposed by Jules Verne. It is a little known fact that the German big guns of World War I (the approximately 100 foot long Big Bertha and a larger gun) that bombarded Paris from a distance of 80+ miles sent their 100+ lb shells as high as 80 – 90 miles above the earth to the very edge of space. The shells had a speed of 1 mile per second as they emerged from the gun's barrel. The German big guns were exceeded by the guns developed in the HARP program led by Jerry Bull 1960's.[11]

Space guns are interesting in the light of the fact that altitudes of about 100 miles are viewed as "Near Space" and that today many space satellites circle the earth at 200+ mile altitudes in "low earth orbit".

---

drinking water by a **factor** of 30,000. Perchlorates appear in 93% of US lettuce and milk, and 97% of US breast milk. Senator Feinstein (CA) has called this situation the "US Rocket Fuel Pollution Scandal" and pressed for strong congressional action.

[11] I am grateful to Dr. Mitat A. Birkan, Program Manager, Space Propulsion and Power, AFOSR/NA for providing this information as well as other details discussed later on chemical space guns.

Space guns can put objects into Near Space. The price to shoot one kilogram (2.2 pounds) at a muzzle speed of 1.6 km/sec is about one kilogram of the best chemical propellant. In comparison a rocket would use many times more propellant to send a kilogram up eighty miles.

A single or multi-stage rocket, with or without boosters, uses over 99% of its weight as fuel to send a payload into space. The fuel is, for the most part, used to propel itself (the fuel) off the ground at an ever-increasing speed into space. In contrast, the propellant for a gun propels the projectile and a fraction of the propellant between the point of burn of the powder charge and the projectile. Thus a sufficiently large gun could efficiently put a payload up eighty miles into Near Space because the propellant propels the payload, and not fuel or a rocket casing. If the space gun is scaled up to send 500 kg payloads into space then we have an effective mechanism to send large amounts of material into space in 500 kg chunks using about 500 kg of propellant per payload.[12]

When the payload reaches eighty km or so then there are two possibilities. The payload might contain a small rocket to put it into a higher earth orbit or to send it to a space station. Or there could be a "scooper" vehicle circling the earth that could scoop up the payload and deliver it to a space station at a higher altitude.[13]

The fabrication of space guns is well within our technology. The major drawback to space guns is the deterioration of the wall of the gun barrel with repeated use. However, the walls can be resurfaced – presumably at a much lower cost than replacing a throwaway rocket or refurbishing a used space shuttle. The other drawback is the design and construction of scooper vehicles. This does not appear to be a significant problem since constantly circling vehicles using atmospheric bounce were designed in the 1950's by NASA although never built.

---

[12] Dr. Mitat points out that the amount of propellant increases rapidly with muzzle velocity and that at over 2 km/s the required propellant mass is more than three times as large as the payload.

[13] The concept of a vehicle constantly circling the earth at a range of altitudes and "bouncing" off the earth's atmosphere to reach higher altitudes was developed by German scientists in World War II and studied by American scientists after the war. This type of vehicle could be used as a scooper vehicle for payloads shot into Near Space.

3.4 Rail Guns

A *rail gun* is an electrical device that accelerates a conducting projectile along a pair of metal rails. Railguns have two sliding or rolling contacts that enable a large electric current to pass through the projectile. The rails generate a very strong magnetic fields that causes the projectile conducting the current to be rapidly accelerated to a high speed.

Chemical propellant guns have been created that have reached speeds of 2.6 km/sec. Rail guns have accelerated projectiles to speeds over 6 km/sec in laboratory experiments. Thus a rail gun track running up the side of a high mountain can accelerate a payload to a speed nearly half the escape velocity of the earth (11.2 km/sec).[14] In addition a rail gun does not have the problem of hot gases, corrosion, and barrel deterioration of chemical space guns although rail erosion is a problem. Replacing rails is generally easier than repairing gun barrels.

They also do have the problem of requiring sizeable amounts of electricity. Various laboratories around the world are engaged in rail gun research.[15]

If rail guns prove feasible and economical they could provide a means to send 100 kg (or greater) payloads into space for pick up by a scooper vehicle.

3.5 Conclusion

Rockets, space guns, and rail guns offer mechanisms to send materials in bulk into space as payloads that can be assembled into a large space station or set of space stations.

These space stations can, in turn, be the terminals for travel back and forth to the moon or planets.

---

14 In contrast the escape velocity from the surface of the moon is 2.4 km/sec.
15 One interesting study of the use of rail guns to assist rocket acceleration is described in Uranga et al "Rocket Performance Analysis Using Electrodynamic Launch Assist", Proceedings of the 43rd AIAA Aerospace Sciences Meeting (January, 2005, Reno, Nevada).

4. Earth Orbit to Moon Orbit and Moon Base

4.1 Moonbase Development Phase

After the establishment of a sizeable, perhaps self-sustaining, space station with major stores of rocket fuel, an organized colonization of the moon becomes possible using a combination of conventional rockets and nuclear rockets.

One possibility would be for a chemical rocket to be assembled at the space station and linked to a nuclear rocket also assembled at the space station. The chemical rocket would carry both itself and the nuclear rocket to the vicinity of the moon – in particular, to just beyond the point where the earth's gravity and the moon's gravity are equal and opposite. Then the nuclear rocket could detach and safely use its nuclear power to establish an orbit around the moon. The nuclear rocket would be beyond the earth's pull and thus the danger of a crash of a nuclear rocket with the reactor operating on earth would be avoided.

The chemical rocket could continue on to land on the moon and be recycled into part of a moonbase. The combination of many rockets suitably modified and joined could form a sizeable initial moon installation. At a later stage the moon's crust could be mined, and hydrogen, oxygen, water, iron, titanium, aluminum uranium[16] and other metals could be refined and used for construction purposes, and to make nuclear spaceships that could take advantage of the moon's weaker gravity compared to the earth.

The nuclear rockets orbiting the moon could form a flotilla of ships to establish transportation to Mars and beyond.

---

[16] The Japanese Kaguya spacecraft named SELENE for "Selenological and Engineering Explorer" has detected thorium, potassium, oxygen, magnesium, silicon, calcium, titanium, iron and *uranium* in the moon's crust since its launch in 2007.

4.2 Moon Colonization

The moon has water, metals and other materials, as well as abundant solar energy to make a moon colony a viable option. The development of mining and manufacturing industries on the moon to produce spacecraft would eventually be more economical than transporting them from earth with an escape velocity almost triple that of the moon.

Since there are many studies and articles on the details of various development proposals we will not discuss a detailed plan here. The reader can find possible plans elsewhere.

4.3 Mature Nuclear Transportation

The engineering of nuclear rockets can be expected to follow a pattern similar to the improvements in airplanes and jets seen in the twentieth century. Through trial and error, and the analysis of mishaps, nuclear rockets will be improved to the point where they will eventually take over transportation between earth space stations and the moon and planets.

The decided advantages of nuclear rockets in terms of speed and economy will eventually make them the preferred method of travel in the solar system.

5. Mars Colonization

5.1 Transportation to Mars

Although there are many ingenious proposals for travel to Mars and other planets the most economical, fastest and safe mode of transportation will be nuclear rockets after nuclear rocket technology has matured through use in space. There is no substitute for the practical experience, and the innovation that it stimulates, through the actual use of nuclear rockets. Examples of similar growth in the efficiency and sophistication of vehicles are the history of automobile and airplane development.

Nuclear rocket traffic to Mars would initially originate from moon orbit until rocket safety had reached a point where trips to Mars could originate from earth orbit.

Initially the Martian moons could be used as terminals for traffic from earth – with a shuttle service to the planet's surface.

5.2 Changing the Atmosphere on Mars

The recent discovery of large amounts of water and frozen carbon dioxide on Mars gives hope to the possibility of transforming the Martian atmosphere into a thicker carbon dioxide atmosphere that eventually could evolve to have sufficient oxygen for human needs. There appear to be a number of feasible approaches. One approach might be to explode large hydrogen bombs to vaporize the Martian ice caps and create a thicker atmosphere with carbon dioxide and water vapor. Thinking on a 50 – 100 year time frame the effects of radioactivity would not be of importance. The possibility of a "nuclear winter" resulting from the explosion of the bombs could be reduced by exploding them below ground at a depth sufficient to minimize dust thrown into the atmosphere.

Another approach would be to divert icy comets or meteors using nuclear explosives to bombard Mars. Recently a reservoir of icy comets has been found circling Jupiter. In addition to adding water to the planet they would also serve to heat it.[17]

5.3 Introduction of Plant life

After the generation of a relatively thick and warm atmosphere the introduction of (perhaps genetically modified) plants and eventually animals on a planetary scale would become feasible. The primarily carbon dioxide atmosphere could then be transformed to a breathable atmosphere containing oxygen so that humans could live in a manner similar to that of higher altitude regions of earth.

Then colonization could proceed with the development of business, industry and tourism. Humanity would then have another home world.

---

[17] Recently evidence has surfaced that Mars may have been bombarded by icy comets and meteors in the past that created massive water movements as well as heating the planet to earth-like temperatures for periods lasting up to hundreds of years. See the article: **Environmental Effects of Large Impacts on Mars** by Teresa L. Segura, Owen B. Toon, Anthony Colaprete, and Kevin Zahnle *Science* **298** (Dec. 6, 2002) pages 1977-1980.

6. Venus Colonization – A Long Term Project

6.1 Proposal for a Tri-Planet Homeland for Humanity

The transformation of Venus into a home for mankind also may be possible over a period of thousands of years. In this case we must trap almost 99% of its (primarily) carbon dioxide atmosphere within the rock on the planet's surface creating an earth-like atmospheric pressure and reducing the greenhouse effect (perhaps through a nuclear winter or comet bombardment).

We must also remove sulfur from the atmosphere.[18] Next we must generate an atmosphere containing oxygen and surface water (possibly through bombardment with diverted icy comets as in the case of Mars). The remaining issues: the slow rotation of Venus and the lack of a magnetic field create complications but they can also be overcome by an advanced society.

Some might feel that the investment of vast efforts over a period of perhaps thousands of years is not a reasonable proposal. However, the reward for such an effort would be another earth at a relatively close distance. Thus humanity would benefit from an enormous expansion of its living space. The accomplishment of this goal would also have the beneficial effect of vastly increasing the scope of human civilization.

---

[18] The early earth also had very little oxygen and enormous amounts of methane and sulfur in the atmosphere that was eventually trapped in rocks forming sulfites. See the articles: Farquhar, J., K.D. McKeegan and M.H. Thiemens, **Mass-independent sulfur of inclusions in diamond and sulfur recycling on early Earth** *Science* **298** (Dec. 20, 2002) pages 2369-2372; Habicht, K.S., and D.E. Canfield, **Calibration of Sulfate Levels in the Archean Ocean** *Science* **298** (Dec. 20, 2002) pages 2372-2374; Wiechert, U.H., **Earth's Early Atmosphere** *Science* **298** (Dec. 20, 2002) pages 2341-2342.

6.2 Interplanetary Civilization

One of the major points of the work of Arnold Toynbee, and other students of civilization, is that a civilization grows and matures through meeting great challenges. As the Egyptians conquered the Nile Valley to produce a great civilization, Mankind has the opportunity to conquer the nearby planets to achieve a new level of civilization.

The accomplishment of these projects will give humanity three earth-like homes. If large extraterrestrial human colonies are established then a human space civilization will develop. *The size of the challenge confronting this future civilization is such that a successful response will undoubtedly create a civilization that may be an order of magnitude above current human civilizations.* The mushroom ring of civilization will then have expanded to the planets with a clear view towards the stars.

The United States, the European nations, Russia, China and India are the only entities with sufficient resources to lead a major move into space. Rather than devote resources to war (and preparations for war) they should create a major presence in space.

Militarism will waste the resources of the world just as Trajan wasted the resources of Rome in the expansion of its frontiers. EuroRussoSinoIndianAmerican resources could be better devoted to another purpose: creating major space settlements—that would be to the world as the United States was to Great Britain—the offspring that helped to save the parent (in two world wars.)

7. Solar System Colonization

Not content to colonize the inner planets it is reasonable to consider the development of settlements in the asteroid belt, the moons of the outer planets, and perhaps even as far as the Oort Belt.

Initially such settlements would be for scientific purposes. One could imagine, for example, an array of radio and optical telescopes ranging from one edge of the solar system to the other. If the components of the array were properly synchronized as we do in telescope arrays on earth at present we would have a giant eye on the universe that would help us detect cosmic phenomena, and particularly, detect solar systems around other stars with earthlike planets. This telescope array would then help determine the target stars to visit as we begin to explore the stars.

Eventually colonies in the outer solar system might be developed for mining, industry and other purposes. The low intensity of sunlight in those regions would require artificial light for health and other reasons.

8. Superluminal Motion

8.1 Some Attempts at Superluminal Travel

To travel to the stars one needs a means of transportation that is fast, economical, capable of holding large cargoes, and within the capabilities of current technology and world resources. Most of the proposed methods of transportation do not meet all or most of these criteria.

Consequently, a new means of starship propulsion must be discovered. One proposal that has drawn significant theoretical interest is the Alcubierre[19] drive named after the originator of the concept. This approach is based on creating a distortion in space: a local expansion of space-time behind the starship and a contraction of space-time in front of it. The result is *faster than light travel – superluminal motion.* Unfortunately this proposal has been shown[20] to require the presence of exotic matter such as tachyons *in massive quantities* since the force of gravity is weak. Thus it does not appear to be feasible. None of the gravity-based approaches are feasible.

8.2 Tachyons and Superluminal Travel

In some previous studies we showed that 1) particles entering a black hole acquire a superluminal speed, and also 2) that one can derive the Standard Model of Elementary Particles if one assumes neutrinos and down-type quarks are tachyons.[21] Since neutrinos have an extremely small mass it is not yet possible to verify that they are tachyons. Since quarks are confined within protons, neutrons, and so on, the nature of d-

---

[19] M. Alcubierre, Class Quantum Grav **11**, L73 (1994); arXiv:gr-qc/0009013 (2000).
[20] R. J. Low, arXiv:gr-qc/9812067 (1998) and references therein.
[21] Blaha (2006), (2007a), (2007b), (2008).

type quarks is also not known with certainty. However the difficulties of interpretation of the spin characteristics of deep inelastic electron-nucleon scattering suggest quarks are not normal spin ½ particles. These discrepancies may be explainable if d-type quarks are tachyons. Tachyons have different spin characteristics than normal spin ½ particles.

Given the possibility of tachyons in nature, and remembering the history of black holes which were predicted many years ago and only found recently, it is reasonable to think that tachyons will eventually be found – perhaps in studies of the quark-gluon plasmas that are now being produced at new accelerators in Au-Au and other heavy atom scattering. We note again that particles inside black holes are tachyons. So the question devolves to finding tachyons outside of black holes.

If we can create tachyons outside of black holes with a sufficient density to modify the space-time metric tensor $g_{\mu\nu}$ then starships traveling at superluminal speeds become possible.

Since tachyons are a form of exotic matter Alcubierre's drive could become feasible if they could be created in truly massive quantities of the order of many solar masses. This possibility seems to be beyond the capabilities of any foreseeable civilization.

So we will consider an alternate possibility in chapter 10 which uses tachyons in the form of a quark-gluon plasma as the starship propellant. This new approach appears to be feasible using some extensions of current technology and does not require massive resources beyond those currently available. In the following sections we will describe some important features of tachyonic motion.

8.3 Tachyons

Tachyon particles, particles traveling at a speed faster than light, were first proposed[22] in the 1960's. The concept of a superluminal particle was based on the possibility that the speed of light is not an absolute limit but a type of "horizon" beyond which sub-light particles could not go in our normal, almost flat space-time. The concept of superluminal particles was recognized as not creating paradoxes but

---

[22] S. Tanaka, Prog. Theoret. Phys. (Kyoto) **24**, 171 (1960); O. M. P. Bilaniuk, V. K. Deshpanda and E. C. G.; Sudarshan, Am J. Phys. **30**, 718 (1962); G. Feinberg, Phys. Rev. **159**, 1089 (1967); M. E. Arons & E. C. G. Sudarshan, Phys. Rev. **173**, 1622 (1968). Feinberg coined the term tachyon in his 1967 paper.

early work on the quantum theory of such particles was not able to establish a successful free quantum theory of tachyons. Blaha (2007a) contains the first successful quantum theory of free tachyons.

Einstein said that he was led to the Special Theory of Relativity by considering an observer traveling at the speed of light and viewing an electromagnetic wave traveling at the same speed. The electromagnetic wave would appear to be static – not having the oscillatory nature of a wave – contrary to the theory of electromagnetism. He concluded that no observer (or particle that might "hold" an observer) could travel at the speed of light and thus all particles must travel at a speed below the speed of light. In fact, he actually proved that particles with mass cannot travel at exactly the speed of light.

He did not consider the possibility that particles might exist that travel at a speed greater than the speed of light. Blaha (2007a) considered exactly that case and found that particles traveling faster than the speed of light would "see" electromagnetic waves traveling at precisely the speed of light and oscillate according to the laws of electromagnetism.[23] Thus Einstein's thought experiment only ruled out the possibility of massive particles traveling at exactly the speed of light.

We will therefore assume tachyons can exist in nature but remain to be found experimentally outside of black holes.

8.4 Extension of Special Relativity to Include Faster-than-Light Particles

The Lorentz group, the essence of Special Relativity, relates the coordinates of an event in two coordinate systems that differ by a relative velocity whose magnitude is less than the speed of light. We imagine an observer on one "particle" (the "lab" observer) observing at event at time t at the position (x, y, z). Another observer on another "particle" traveling at a speed v in the x direction relative to the first observer (as depicted in Fig. 8.2) observes the same event at time t' at the position (x', y', z'). The relation between the coordinates of the two observers is given by the Lorentz transformation

$$t' = \gamma(t - \beta x/c) \qquad (8.1)$$
$$x' = \gamma(x - \beta ct)$$

---

[23] Blaha (2007a) p. 12 and earlier work.

$$y' = y$$
$$z' = z$$

where $\beta = v/c$, c is the speed of light, and $\gamma = (1 - \beta^2)^{-\frac{1}{2}}$.

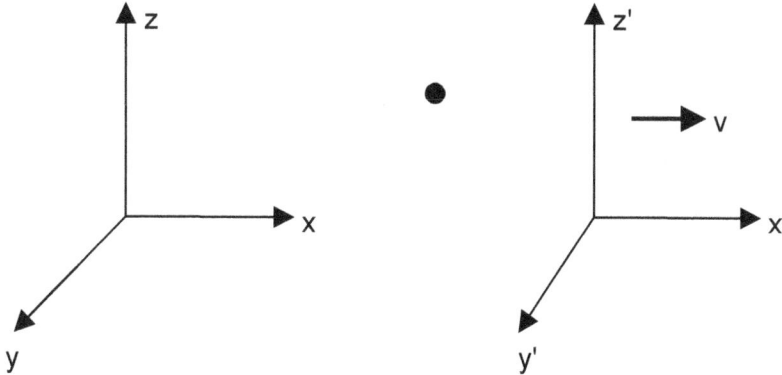

Figure 8.2. Two coordinate systems having a relative speed v in the x direction. The black circle represents an event. The unprimed coordinate system is the "lab" system while the primed coordinate system is the system of an observer "on a particle" moving with speed v along the parallel x and x' axes.

If β is less than one (sublight speed) then the coordinates of an event are related by eq. 8.1, and specify the time and location of an event from the viewpoint of an observer at rest in each coordinate system. If β is greater than one (superluminal speed) then the coordinates of an event are still related by eq. 8.1, and still relate the time and location of an event from the viewpoint of an observer at rest in each coordinate system. However t' and x' are now imaginary numbers since γ is an imaginary number.

How can we physically understand this state of affairs? Well, if we consider the realities of the observer in the "primed" coordinate system it is clear he/she will measure the x' distance with a ruler that measures real numbers, and he/she will measure time t' with a clock that measures real numbers. So the imaginary values of x' and t' in eq. 8.1 only appear in the relation between the coordinate systems. If we denote the actual values measured by the primed coordinate system observer as t_r' and x_r' then eq. 8.1 for a superluminal relative speed v becomes

$$t_r' = \gamma_s(t - \beta x/c) \qquad (8.2)$$
$$x_r' = \gamma_s(x - \beta ct)$$
$$y' = y$$
$$z' = z$$

where

$$\gamma_s = i\gamma = (\beta^2 - 1)^{-\frac{1}{2}} \qquad (8.3)$$

is a real positive number for $\beta > 1$.

The 4-vector inner product is invariant for $\beta > 1$ for t' and x' as given by eq. 8.1:

$$c^2 t^2 - x^2 - y^2 - z^2 = c^2 t'^2 - x'^2 - y'^2 - z'^2 \qquad (8.4)$$

just as it is for $\beta < 1$. A purist might object and say that the 4-vector inner product should be invariant for the physically measured quantities in each coordinate system. However times and lengths are always measured as real numbers and so the fault, if there is one, is that rulers and clocks don't display values with the imaginary number i.

The simplified discussion given above of the Lorentz transformation for sublight and superluminal relative speeds in the same direction is expanded to a complete discussion of the complex Lorentz group and relative velocities in any direction in Blaha (2007a) and subsequent books in 2007 and 2008.[24]

8.5 Time and Space Contraction and Dilation

In ordinary Lorentz transformations a moving ruler will appear to be shorter in the direction of its motion when measured in another reference frame. This phenomenon is called *Lorentz contraction*. In ordinary Lorentz transformations time intervals will appear to be longer when measured in another reference frame. This phenomenon is called *time dilation.*

---

[24] And "left-handed" complex Lorentz transformation are shown to lead to parity violation, as well as other major features, in a derivation of the form of the Standard Model of Elementary Particles.

In superluminal transformations contraction and dilation are more complicated as we will see in this section.

Superluminal Length Dilation/Contraction

In the case of a superluminal transformation we find *superluminal length contraction or dilation* can occur depending on the relative velocity. Consider the case of the transformation of eq. 8.1 above, which relates the primed reference frame traveling at speed v in the positive x direction to the unprimed reference frame. A ruler perpendicular to the x-axis will have the same length in both reference frames if its endpoints are simultaneously measured – perhaps by photographing it. The y and z equations in eqs. 8.1 specify this fact.

If the ruler is at rest in the primed reference frame and parallel to the x' axis, then a simultaneous measurement of its endpoints at the same time t_0 by an observer in the unprimed reference frame (perhaps by photographing it) will reveal either *length contraction and dilation* depending on the value of β. If the length is $L' = x'_2 - x'_1$ in the primed frame and $L = x_2 - x_1$ in the unprimed frame, then the equations:

$$x'_1 = \gamma_s(x_1 - \beta c t_0) \tag{8.5}$$
$$x'_2 = \gamma_s(x_2 - \beta c t_0) \tag{8.6}$$

imply

$$L' = \gamma_s L = (\beta^2 - 1)^{-\frac{1}{2}} L \tag{8.7}$$

Thus we have three cases:

Case 1: $\beta \in <1, \sqrt{2}>$: $\qquad\qquad$ L < L' $\qquad$ Contraction $\qquad$ (8.8)

Case 2: $\beta = \sqrt{2}$: $\qquad\qquad$ L = L' $\qquad$ Equality $\qquad$ (8.9)

Case 3: $\beta \in <\sqrt{2}, \infty>$: $\qquad\qquad$ L > L' $\qquad$ Dilation $\qquad$ (8.10)

Thus $\beta = v/c = \sqrt{2}$ marks the point of change from the Lorentz contraction of lengths to dilation of lengths. This feature of superluminal motion, first noted in Blaha (2007a), has no counterpart in sublight motion.

The effect of the change at $\beta = \sqrt{2}$ on a starship is startling. Imagine the primed coordinate system is that of the starship and the unprimed coordinate system is that of the earth. (The motion of the earth is small and can be neglected relative to the high speed of the starship.) If the starship velocity moving on a straight line away from the earth is such that β is between 1 and $\sqrt{2}$ then a yardstick on the starship will appear to be shorter in the earth coordinate system – Lorentz contraction. If the starship velocity is such that β is greater than $\sqrt{2}$ then a yardstick on the starship will appear to be longer in the earth coordinate system – dilation. More interestingly, if the starship travels 1 light year in its coordinate system it will actually have traveled more than 1 light year in the earth's coordinate system – possibly *much* more than 1 light year in the earth's coordinate system – if the starship's speed has a β that is much greater than $\sqrt{2}$.

Superluminal Time Contraction/Dilation

In the case of a superluminal transformation *superluminal time contraction* is a possibility.[25] Consider again the case of the transformation of eq. 8.1 relating the primed reference frame traveling at speed v in the positive x direction to the unprimed reference frame. Consider the time interval between two events occurring at the same point x'_0 in the primed reference frame. From the viewpoint of an observer in the unprimed frame the events take place at different points x_1 and x_2. If the time interval is $T' = t'_2 - t'_1$ in the primed frame and $T = t_2 - t_1$ in the unprimed frame, then the inverse transformation to eq. 8.1 gives:

$$t_1 = \gamma_s(t'_1 + \beta x'_0/c) \qquad (8.11)$$
$$t_2 = \gamma_s(t'_2 + \beta x'_0/c) \qquad (8.12)$$

and implies

$$T = \gamma_s T' = (\beta^2 - 1)^{-\frac{1}{2}} T' \qquad (8.13)$$

Again there are three cases:

Case 1: $\beta \in <1, \sqrt{2}>$: $\qquad\qquad$ T > T' $\qquad$ Dilation $\qquad\qquad$ (8.14)

Case 2: $\beta = \sqrt{2}$: $\qquad\qquad\qquad$ T = T' $\qquad$ Equality $\qquad\qquad$ (8.15)

Case 3: $\beta \in <\sqrt{2}, \infty>$: $\qquad\qquad$ T < T' $\qquad$ Contraction $\qquad$ (8.16)

---

[25] Blaha (2007a).

The time interval in the unprimed frame can be less than, equal to, or greater than the time interval in the primed frame when the events take place at the same spatial point.

Thus superluminal transformations are more complex than sublight Lorentz transformations (which only have time dilation) with respect to time dilation and contraction.

The effect of the time change at $\beta = \sqrt{2}$ on a starship is also startling. Again imagine the primed coordinate system is that of the starship and the unprimed coordinate system is that of the earth. If the starship velocity moving on a straight line away from the earth is such that β is between 1 and $\sqrt{2}$ then a time interval on the starship will appear to be longer in the earth coordinate system – time dilation. Or, stating it otherwise, time intervals on the starship will appear to be shorter than they appear on earth. A person on a starship would thus age more slowly from the point of view of a person on earth.

If the starship velocity is such that β is greater than $\sqrt{2}$ then a time interval on the starship will appear to be shorter in the earth coordinate system – time contraction. If the starship travels for 1 year in its coordinate system it will actually have traveled less than a year in the earth's coordinate system – possibly much less than a year in the earth's coordinate system – if the starship's speed has a β that is much greater than $\sqrt{2}$. Thus people on the starship would appear to age more quickly than on earth.

Combined Effect of Space and Time Dilation and Contraction

If we take account of the combined effects of space and time dilation and contraction respectively (eqs. 8.10 and 8.13) we find that

$$L/T = (\beta^2 - 1)L'/T' \tag{8.17}$$

Now imagine a trip from earth to a star at the constant speed v of distance L and travel time T (neglecting initial acceleration and destination deceleration) and assume that v is such that β is greater than $\sqrt{2}$.[26] Then the effective speed as viewed from earth will be more than the effective speed as viewed in the starship. As a result the starship crew will believe it arrived at its destination more slowly than people on earth

---

[26] The effects of β greater than $\sqrt{2}$ were first noted in Blaha (2007a).

would think. So the combined effect of space and time dilation and contraction for β greater than $\sqrt{2}$ is to make the starship crew think it traveled more slowly to its destination. If β is much greater than $\sqrt{2}$ then the starship crew will think it arrived much more slowly then the earth based observers – especially due the appearance of the square of β in eq. 8.17.

8.6 Mass of a Starship – Tachyonic!

Lorentz transformations apply to the momentum of particles (and clumps of particles) as well as to the coordinates of particles. So we can take the simple case considered earlier of eq. 8.1 and relate the momentum of the starship in its (primed) coordinate system to the momentum of the starship as seen from earth (the unprimed coordinate system) using the same transformation as in eq. 8.1.

$$E' = \gamma(E - \beta c p_x) \qquad (8.18)$$
$$p'_x = \gamma(p_x - \beta E/c)$$
$$p'_y = p_y$$
$$p'_z = p_z$$

where $\beta = v/c$, c is the speed of light, and $\gamma = (1 - \beta^2)^{-\frac{1}{2}}$. E and E' are the energies in the respective coordinate systems. And the spatial momenta are (p_x, p_y, p_z) and (p'_x, p'_y, p'_z) in the coordinate systems.

If we now assume β is greater than one then we obtain the superluminal transformation of momenta corresponding to eq. 8.2:

$$E' = \gamma_s(E - \beta c p_x) \qquad (8.19)$$
$$p'_x = \gamma_s(p_x - \beta E/c)$$
$$p'_y = p_y$$
$$p'_z = p_z$$

where
$$\gamma_s = i\gamma = (\beta^2 - 1)^{-\frac{1}{2}} \qquad (8.3)$$

In the example we are considering $p_y = p_z = 0$ and in the primed frame where the starship is at rest $p'_x = p'_y = p'_z = 0$ and $E' = M'c^2$ the mass of the starship in its rest frame coordinate system – the primed frame.

　　　　The momenta of a particle in a coordinate system are defined to have the form:

$$p^0 = E/c \qquad \text{and} \qquad p^i = mv^i\gamma/c \qquad (8.20)$$

where i = 1, 2, 3 corresponding to the x, y, and z components of the velocity of the particle. The energy $E = \gamma mc^2$ and $\gamma = (1 - \beta^2)^{-\frac{1}{2}}$ with m the mass of the particle in the coordinate system.

　　　　Therefore the 4-vector inner product of the momenta (corresponding to eq. 8.4 for the coordinates) of a starship in its own rest frame is

$$p'^{0\,2} - \Sigma\, p'^{i\,2} = (E'/c)^2 = M'^2c^2 \qquad (8.21)$$

where M' is the mass of the starship in the primed coordinate system.
Now E and p_x have the form given in eq. 8.20 with mass M in the unprimed coordinate system. Substituting for E' using the transformation law eq. 8.19 and

$$E = Mc^2\gamma \qquad\qquad p_x = Mv\gamma \qquad (8.22)$$

yields

$$(E'/c)^2 = [\gamma_s(E - \beta c p_x)/c]^{\,2} = M^2c^2\gamma_s^2\gamma^2(1 - \beta^2)^2 = -M^2c^2$$

and thus

$$M^2 = -M'^2 \qquad (8.23)$$

Thus a particle, and by extension an entire starship, is tachyonic from the viewpoint of the earth's coordinate system. *Tachyonic Starships!*

9. Generalization of General Relativity to Include Superluminal Coordinates

This section is included for completeness to show that our generalization of Special Relativity to include superluminal transformations can be placed within the framework of a generalization of General Relativity to complex coordinates. Some readers may choose to skip this chapter since it is not directly pertinent to our starship design proposal or to travel to the stars. Since space is almost flat outside of large masses such as black holes and stars, starship travel will be based on the flatness of most of space within our galaxy and between galaxies.

9.1 Complex General Relativity

The generalization of Lorentz transformations – Special Relativistic transformations – to superluminal transformations requires General Relativity to be generalized to include complex coordinates. The superluminal transformation in the special case where the spatial axes of the coordinate systems are parallel and the relative velocity is in the x direction is given by eq. 8.2. x' and t' are imaginary in this case. In the general case where the relative velocity is in an arbitrary direction the primed coordinate system coordinates are complex in general. Yet the invariant inner product of superluminal 4-vectors has the usual form of Lorentz transformations, namely,

$$x{\cdot}y = c^2x^0y^0 - \Sigma\, x^iy^i \qquad (9.1)$$

and not

$$x^*{\cdot}y = c^2x^{0*}y^0 - \Sigma\, x^{i*}y^i \qquad \text{NO} \qquad (9.2)$$

In General Relativity the invariant infinitesimal proper time has the form

$$d\tau^2 = g_{\mu\nu}dx^\mu dx^\nu \qquad\qquad (9.3)$$

where $g_{\mu\nu}$ is the metric tensor. Coordinates and infinitesimal coordinate displacements dx^μ are real. It is invariant under real general coordinate transformations:

$$x^\mu = f^\mu(x'^0, x'^1, x'^2, x'^3) \qquad\qquad (9.4)$$

for $\mu = 0, 1, 2, 3$. The dx^μ are vectors:

$$dx^\mu = dx'^\nu \, \partial f^\mu(x'^0, x'^1, x'^2, x'^3)/\partial x'^\nu \qquad\qquad (9.5)$$

or, more simply,

$$dx^\mu = dx'^\nu \, \partial x^\mu/\partial x'^\nu \qquad\qquad (9.6)$$

When we generalize General Relativity to complex coordinates the issue that immediately arises is the form of the proper time infinitesimal. The usual choice is to define the Complex General Relativity proper time as

$$d\tau^2 = g_{\mu\nu}dx^{\mu}*dx^\nu \qquad\qquad \text{NO} \qquad\qquad (9.7)$$

However that would be inconsistent with the form of $d\tau^2$ for the combined theory of Special Relativity and superluminal transformations that we have called L_C in prior books.[27]

We therefore define the invariant interval for our Complex General Relativity as eq. 9.3 where the metric and coordinates can be complex and the metric is an holomorphic function of the coordinates.[28] Coordinate transformations have the form of eq. 9.4 with the proviso that f^μ are continuous, holomorphic functions.

A major benefit follows from this definition of complex General Relativity. As p. 19 of Blaha (2004) points out:

> Since the definition of a complex partial derivative is formally identical to the partial derivative with respect to a real variable we can make the following observation.

---

[27] Blaha (2008) and earlier books.
[28] Blaha (2004) develops this theory in detail.

Observation: *All the formal rules of complex partial differentiation are the same as those of real partial differentiation. Thus the differentiation of sums, products, quotients and so on are formally the same as in the real variable case. In particular, a complex partial derivative will coincide with the corresponding real partial derivative on the real axis (i.e. for X^ν real).*

Thus we are saved the labor of recalculating the equations of Riemannian geometry for the complex case.

The reader can read a detailed account of analytic complex General Relativity in Blaha (2004) including its Schwarzschild solution and Robertson-Walker solution as well as a quantum theory of the Big Bang.

10. Quark Drive Starship

10.1 Starship Approaches that Don't Work

Many proposals have been made for travel to the stars.[29] These proposals fall into one or both of two categories: 1) they require resources beyond the present or foreseeable resources of humanity or planet earth; 2) they are not feasible due to physics, as we know it, such as the weakness of the force of gravity. As a result they do not seem to be viable either now or in the future.

One intriguing approach is that of M. Alcubierre[30] who proposes to have a starship "ride a kink in the gravitation field" moving at speeds much greater than the speed of light. As Alcubierre noted, and R. J. Low showed, this approach is not feasible. The creation of such a "kink" appears to be impossible without the use of an *enormous* mass of exotic matter due to the weakness of the force of gravity in addition to the reason adduced by Low.

10.2 Chapter Summary

In this chapter we will proceed in a deliberate way to develop the concepts that will lead to a plausible starship design.

1. First we will review the case of a rocket accelerating indefinitely and show that it can never reach the speed of light due to the dynamical restrictions of Special Relativity.
2. Then we will consider the possibility of complex velocities – velocities with x, y, and z components that are not real numbers but are complex numbers.

---

[29] Mallove and Matloff (1989) describe several proposals.
[30] M. Alcubierre, Class. Quantum Grav. **11**, L73 (1994); R. J. Low, Class. Quantum Grav. **16**, 543 (1999); M. Alcubierre, arXiv:gr-qc/0009013 (2000); and references therein.

3. We then show that the speed of light can be exceeded using starships traveling at complex velocities.
4. We will then point out that a recent derivation of the major features of the Standard Model of Elementary Particles attributes complex velocities to quarks and strong interaction gluons.
5. Next we consider quark-gluon plasmas which have been created at large particle accelerators by colliding gold ions and other heavy ions at high energy.
6. If the particles in these plasmas have complex velocities and if plasma regions can be enlarged to macroscopic sizes then it becomes possible to consider using quark plasmas to drive starships with complex velocities that can exceed the speed of light.
7. We will then consider one simple tachyonic starship design based on this approach.

The chapter assumes the Standard Model formulation developed in Blaha (2008) is correct and that macroscopic quark-gluon plasmas are possible. In a sense this second assumption is reminiscent of the development of nuclear energy. First Einstein showed that mass could be converted into energy ($E = mc^2$). Then in the 1930's nuclear fission of individual atoms was found in laboratories (as we today have seen quark-gluon plasmas created by the high energy collision of two gold atoms). A few years later nuclear fission on a macroscopic scale was performed in nuclear reactors and atomic bombs. The analogous creation of macroscopic quark-gluon plasmas remains to be done.

10.3 An Accelerating Rocket in Special Relativity

In this section we consider the case of a rocket undergoing constant acceleration in the rocket's coordinate system (rest frame) which we will identify with the primed coordinate system of eqs. 8.1. The unprimed coordinate system (the "lab" system) we will identify as the earth.

In the rocket's rest frame we will assume that the rocket is experiencing a constant force (thrust) of magnitude g in the positive x direction:

$$\mathbf{F'} = g\hat{\mathbf{x}} \qquad (10.1)$$

The fourth component of the force (since force is a Lorentz 4-vector) is zero in the rocket's rest frame:

$$F'^0 = 0 \qquad (10.2)$$

Applying the inverse of the Lorentz transformation eq. 8.2 we find the force in the earth rest frame is

$$F^0 = \gamma(F'^0 + \beta F'^x/c) = \gamma\beta F'^x/c = \gamma vg/c^2 \qquad (10.3)$$
$$F^x = \gamma(F'^x + \beta cF'^0) = \gamma F'^x = \gamma g$$
$$F^y = F^z = 0$$

where $\beta = v/c$, c is the speed of light, and $\gamma = (1 - \beta^2)^{-\frac{1}{2}}$ as before. We use the superscripts x, y, and z to identify the components of the spatial force. The relativistic expression for the momentum of an object of mass m is

$$\mathbf{p} = \gamma m\mathbf{v} \qquad (10.4)$$

and the dynamical equation for the motion of an object is

$$d\mathbf{p}/dt = \mathbf{F} \qquad (10.5)$$

in the "earth" coordinate system resulting in

$$dp^x/dt = \gamma g \qquad (10.6)$$

with

$$dp^y/dt = dp^z/dt = 0 \qquad (10.7)$$

The differential equation resulting from eq. 10.6 is

$$d(\gamma v)/dt = \gamma g/m \qquad (10.8)$$

which has the solution

$$v = c\{1 - 2/(1 + ((c + v_0)/(c - v_0))\exp[2g(t - t_0)/(mc)])\} \qquad (10.9)$$

where the velocity is v_0 at time t_0. Eq. 10.9 shows that the velocity v will never reach the speed of light c in a finite amount of time.

This simple example illustrates the inability of a massive object to reach the speed of light, or surpass it, in Special Relativity.

10.4 Complex Velocities

The fundamental reason that massive particles cannot attain or surpass the speed of light in Special Relativity is the factor of γ that necessarily appears in the dynamical equation of motion. The singularity of γ for $\beta = \pm 1$ limits the magnitude of velocities to $\beta < 1$.

Up to this point we have considered real quantities: real velocities, real coordinates, and so on. If we extend our considerations to complex velocities then velocities whose real part is greater than the speed of light becomes possible.

Recently Blaha has developed an extension of the Lorentz group that leads to the major features of the Standard Model including parity violation and a differentiation between the leptonic sector and the quark sector. Leptons and ElectroWeak vector bosons have real momenta while quarks and gluons have complex 3-momenta and velocities. (The complex 3-momenta of the quark sector naturally leads to color SU(3).) The simplicity of Blaha's theory, and its direct natural derivation of the Standard Model, strongly argue for its acceptance as a true description of the source and nature of elementary particles. If so, then complex 3-momenta (spatial momenta) are truly a part of nature and we can explore the impact of complex 3-momenta on the question of exceeding the speed of light.

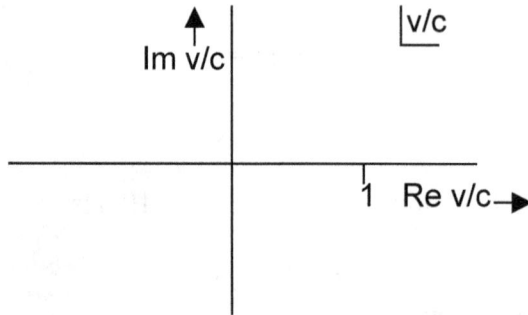

Figure 10.1. The complex velocity plane depicted with complex v divided by the speed of light c, a real number. γ is singular at v/c = 1.

So we will assume that complex 3-momenta are allowed in nature subject to the condition that objects with complex 3-momenta are confined to local regions of space and thus only real momenta are observed in experiments outside of the interaction region of colliding objects. Quark confinement is a manifestation of this concept. Free quarks, which would have complex 3-momenta, are not found. But quarks bound together in elementary particles with real 3-momenta are observed.

10.5 Exceeding the Speed of Light

In section 10.3 we considered the relativistic dynamics of a massive object and showed that under a constant *real* force it could never attain the speed of light. The same would be true if the force were variable.

In this section we will consider the case of a *complex* force that leads to a complex velocity whose real part can exceed the speed of light. Furthermore the velocity and resulting path in space periodically becomes purely real. Thus one can envision traveling at an enormous speed through complex space to a destination with real coordinates. Fig. 10.2 qualitatively depicts the speed of an object from an initial real speed $v_0 = .0001c$ to a real speed peak of 5000c. Higher speeds are accessible. In fact there is no limit on superluminal speeds.

After reaching the desired speed a starship can turn off its engines and "coast" at that speed until it nears its destination. Then it simply runs the starship in "reverse" until it reaches its destination with its original real speed of .0001c (18.6 miles per second) or another sublight speed.

At this point normal nuclear? rocketry can adjust its velocity to that needed to assume orbit around a star or planet. With speeds of 5000c or much more possibly attainable, the entire galaxy as well as nearby galaxies (such as the Magellanic galaxies and Andromeda) becomes accessible to Mankind.

10.5.1 Superluminal Relativistic Dynamics

In this section we will consider a constant propulsive force in a starship's rest frame that drives the starship from a sublight velocity to a superluminal velocity. The key factor in achieving a superluminal speed is evading the singularity in γ at $v/c = 1$. We accomplish this goal by having a complex force – a force with a real and imaginary part – that generates a complex acceleration and thus a complex velocity that "goes around" the singularity in the complex velocity plane. Since the forces, with which we are familiar are real and lead to real accelerations and velocities, we have to find a mechanism to produce such a force. We describe a possible mechanism in sections 10.6 – 10.10 below.

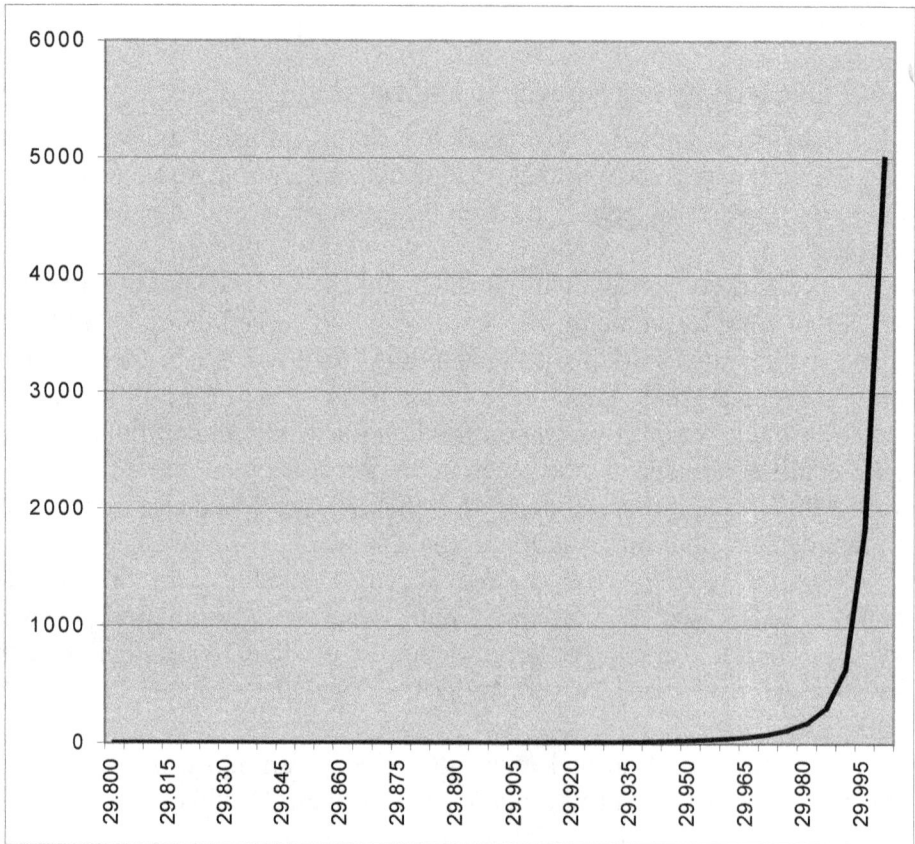

Figure 10.2. An example of the real part of the velocity of a starship on its 29th and 30th day of travel. The dynamics of this case are described in the text. The initial speed of the starship was 0.0001c. Its real speed reaches approximately 5000c and beyond. Speed is measured in units of c; time is measured in days.

For the moment we assume a constant, complex force exists in the rest frame of the starship due to the starship's thrust in the direction of the positive x' (and x) axis. The starship (primed coordinates) and earth (unprimed coordinates) coordinates have parallel axes as in Fig. 8.2. The spatial force is in the positive x direction[31]

$$\mathbf{F'} = g\hat{\mathbf{x}} \qquad (10.10)$$

where g is assumed to now be a complex constant.

The fourth component of the force (since the force is a Lorentz 4-vector) is zero in the rocket's rest frame:

$$F'^0 = 0 \qquad (10.11)$$

Applying the inverse of the Lorentz transformation eq. 8.2 we find the force in the earth rest frame is

$$
\begin{aligned}
F^0 &= \gamma(F'^0 + \beta F'^x/c) = \gamma\beta F'^x/c = \gamma vg/c^2 \qquad (10.12) \\
F^x &= \gamma(F'^x + \beta cF'^0) = \gamma F'^x = \gamma g \\
F^y &= F^z = 0
\end{aligned}
$$

where $\beta = v/c$, c is the speed of light, and $\gamma = (1 - \beta^2)^{-\frac{1}{2}}$ as before. We again use the superscripts x, y, and z to identify the components of the spatial force. The momentum of an object of mass m is

$$\mathbf{p} = \gamma m\mathbf{v} \qquad (10.13)$$

and the dynamical equation of motion is

$$d\mathbf{p}/dt - \mathbf{F} \qquad (10.14)$$

in the "earth" coordinate system resulting in

$$dp^x/dt = \gamma g \qquad (10.15)$$

with

$$dp^y/dt = dp^z/dt = 0 \qquad (10.16)$$

---

[31] The discussion begins by paralleling the previous non-superluminal discussion of section 10.3.

The differential equation resulting from eq. 10.15 is

$$d(\gamma v)/dt = \gamma g/m \qquad (10.17)$$

which has the solution

$$v = c\{1 - 2/(1 + ((c + v_0)/(c - v_0))\exp[2g(t - t_0)/(mc)])\} \qquad (10.18)$$

where the velocity is v_0 at time t_0. Eq. 10.18 has the same form as eq. 10.9 but the complexity of the constant g enables the velocity to exceed the speed of light under certain circumstances that we will explore.

Before doing that we note that eq. 10.18 can easily be integrated to give the distance traveled in the x direction.

$$x = x_0 + (mc/g)\ln((1 - v_0/c + (1 + v_0/c)\exp[2g(t - t_0)/(mc)])/2) - c(t - t_0) \qquad (10.19)$$

or

$$x = x_0 + (mc/g)\ln[(1 - v_0/c)/(1 - v/c)] - c(t - t_0) \qquad (10.20)$$

10.5.2 Determination of Starship Thrust g to Exceed Light Speed

Eqs. 10.18-10.20 apply to sublight and superluminal starship motion. If the thrust of the starship, eq. 10.14, is not real but a complex value along the x axis, then under certain circumstances it can accelerate the starship to complex velocities whose real part can significantly exceed the speed of light. The starship can then travel at this superluminal real speed until it nears its destination. Then it can decelerate to low speed (including zero speed) as it nears its destination. The imaginary part of the velocity causes starship motion in an imaginary direction as well. In the present example the real x axis has an associated imaginary axis which we will call the imaginary x axis. It is really another dimension.[32]

As the starship decelerates the imaginary x distance can be forced to go to zero – resulting in a starship destination location in purely real space and a starship velocity that is also purely real (subsection 10.5.5).

---

[32] Blaha (2004) develops a theory of General and Special Relativity with complex dimensions as have other authors. His theory differs in being holomorphic.

Figs. 10.3 – 10.5 depict the travel of a starship in the complex x plane – the two-dimensional plane consisting of the real and imaginary x axes.

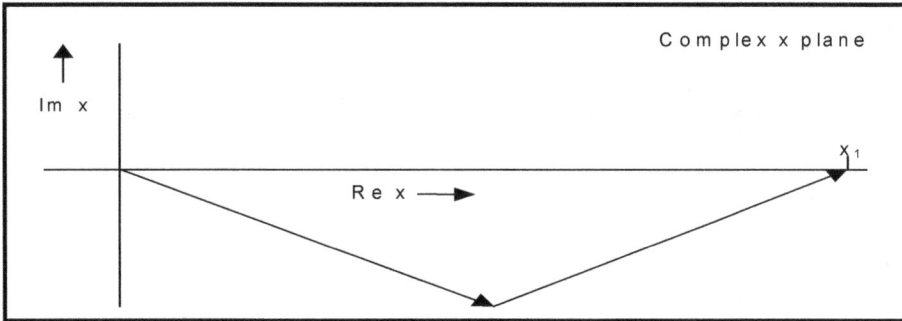

Figure 10.3. Rough depiction of the travel of a starship in the complex x dimensions. The starship starts out in real space at x = 0. While accelerating, the starship is at a complex distance from its origin. After reaching cruising speed it turns off its superluminal engines until near its destination x_1. When nearing its destination it turns the superluminal engines back on, which brings it to its destination at the real distance x_1 at zero imaginary velocity and small real velocity.

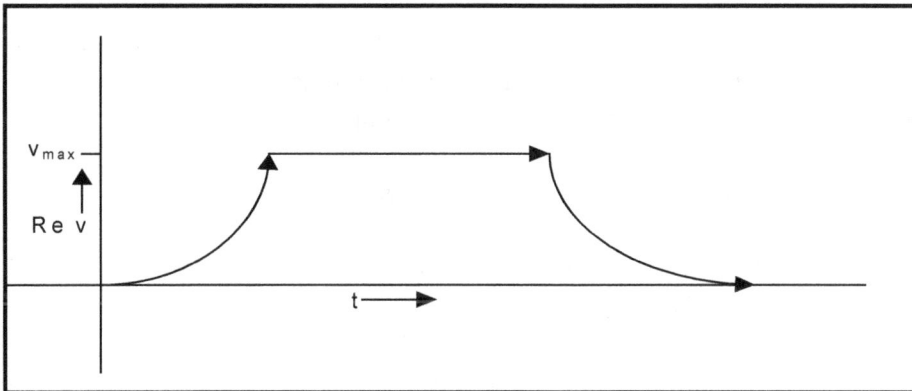

Figure 10.4. Depiction of the real part of a starship speed. There is an acceleration part to a desired maximum speed. Then the starship cruises at that speed until it reaches the vicinity of the destination. Then the starship drive decelerates it to a speed near zero so the starship can enter orbit around a star or planet.

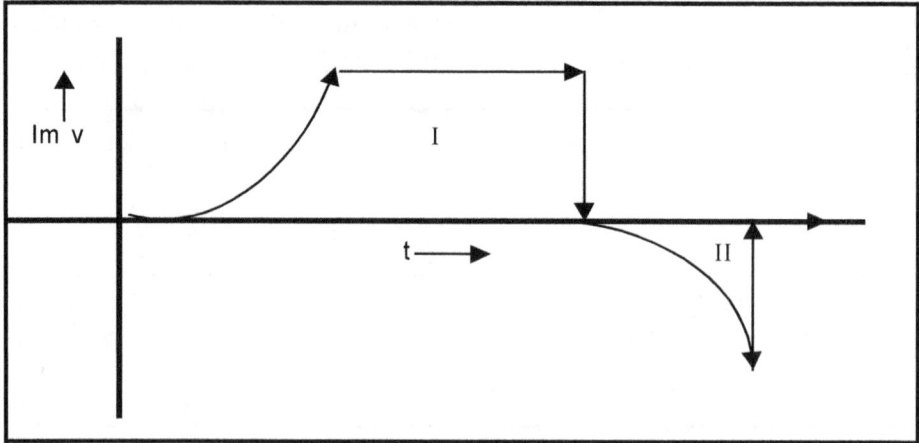

Figure 10.5. Rough depiction of the Imaginary part of a starship speed. There is an acceleration part to a high speed. Then the starship cruises at that speed until it reaches the vicinity of the destination. Then the starship drive decelerates causing the imaginary speed to decrease to zero so the starship has zero imaginary speed at the end of the deceleration. The net imaginary distance traveled is also forced to zero so that the starship ends in real space. The areas of regions I and II are of equal magnitude.

To achieve the motion depicted in Figs. 10.2 – 10.5 the constant force value g must satisfy a special set of conditions. These conditions emerge from a consideration of the denominator of eq. 10.18:

$$1 + ((c + v_0)/(c - v_0))\exp[2g(t - t_0)/(mc)] \qquad (10.21)$$

If this denominator approaches zero then the speed v becomes infinite if g has an appropriate <u>complex</u> value. Let

$$g = g_1 + ig_2 \qquad (10.22)$$

We will describe a mechanism for obtaining a complex thrust, and thus complex g, in sections 10.6 – 10.10 using macroscopic quark-gluon plasmas to drive a starship.

Setting

$$1 + ((c + v_0)/(c - v_0))\exp[2g\Delta t/(mc)] = 0 \qquad (10.23)$$

for some acceleration time interval Δt to infinite velocity (in principle) we find that setting

$$g_2 = n\pi mc/(2\Delta t) \geq 0 \qquad (10.24)$$

and

$$g_1 = (g_2/n\pi)\ln[(c - v_0)/(c + v_0)] \qquad (10.25)$$
$$= (mc/(2\Delta t))\ln[(c - v_0)/(c + v_0)] \leq 0$$

for n an odd, positive integer, enables the real part of the velocity to become infinite at $t - t_0 = \Delta t$. We assume $n = 1$ in the following discussions. Substituting in eq. 10.18 we obtain

$$v = c\{1 - 2/(1 + [(c + v_0)/(c - v_0)]^{1 - (t - t_0)/\Delta t} e^{in\pi(t - t_0)/\Delta t}])\} \qquad (10.26)$$

10.5.3 Acceleration of a Starship from Sublight to Superluminal Speed

We now consider a specific example of the acceleration phase of a starship to get a feeling for a not unrealistic case. First we note that the acceleration of a rocket of mass m with a propellant exhaust speed v_e in the rocket's rest frame is given by

$$dv/dt = (v_e/m)\, dm/dt \qquad (10.27)$$

and thus the constant g of eq. 10.1 is

$$g = mdv/dt = v_e\, dm/dt \qquad (10.28)$$

Since we intend to generate the thrust with a plasma producing an extremely high energy exhaust we will choose the modest value of

$$dm/dt = 1\ gm/sec \qquad (10.29)$$

In one year of travel the amount of mass expelled is approximately 32 metric tons.[33] If we envision a starship of 10,000 metric tons[34] which we

---

[33] The mass of the nuclear (or fusion) fuel needed to generate electricity to power the plasma rings of the starship is another factor but will not be very large compared to the starship's mass.

wish to accelerate to approximately 5,000c starting from a speed of
.0001c (= 18.6 miles per second) within 30 days then the required value
of g is approximately[35]

$$g = -0.3858c + i6060.166c \text{ gm-cm/sec}^2 \qquad (10.30)$$

where $c = 3 \times 10^{10}$ is the numerical value of the speed of light in cgs
units.[36] The consequent exhaust speed is

$$v_e = -0.3858c + i6060.166c \qquad (10.31)$$

which may be feasible using a quark-gluon plasma acceleration
mechanism. We note there is no Special Relativistic limit on imaginary
speed. The required exhaust speed can be reduced if the acceleration
time is lengthened.

Thus the parameters of our starship example in cgs units are

m	10000000000.0000
c	29979245800.0000
g_2	181679202956833.0000
g_1	−11566067091.0259
2g/mc (sec^{-1})	$-7.71604940843838 \times 10^{-11} + 1.21203317901235 \times 10^{-6}i$
2g/mc (day^{-1})	$-6.66666668889076 \times 10^{-6} + 0.104719666666667i$
v_0	0.0001c

The initial starship speed $v_0 = 0.0001c$ or 18.6 miles per second (roughly
three times earth's escape velocity) is well within the capabilities of
nuclear rockets.[37]

The results of calculating the motion from eqs. 10.18 and 10.19
are qualitatively indicated in Figs. 10.4 and 10.5. The velocity increases
slowly until near the 30 day point where singular behavior occurs. The
below table shows the velocity accelerating from $v_0 = 0.0001c$ at the

---

[34] About one-fifth the mass of the ship Queen Elizabeth.

[35] The value of v_e (eq. 10.31) can be decreased by increasing dm/dt. For example if dm/dt = 100
gm/sec then v_e can be decreased by a factor of 100 with the same resulting starship motion.

[36] The mass ejected to generate the thrust to obtain a specific speed will be considered in
Appendix G.

[37] Our starship will have superluminal engines for interstellar travel and nuclear rocket engines
for interplanetary travel and maneuvering. An initial design is discussed later in this chapter.

beginning to its value 29.00006 days later is rather slow compared to the increase in velocity near the 30 day point.

Example of a Starship Acceleration

time (days)	0	29.00006	29.00009	29.999992	29.9999998	29.99999996
Re v	.0001c	30.396c	121.59c	607.9c	5,066c	30,396c
Im v	0	477,465c	1,909,861c	9,549,305c	79,577,538c	4.7745×10^8c

Note the real speed of 5066c at 29.9999998 days (0.017 seconds short of 30 days) and of 4.7745×10^8c at 29.99999996 days (0.0035 seconds short of 30 days). Having achieved the desired real speed the starship's engines turn off and the starship travels at constant speed until near the destination.

At 5066c any of the 100 or so known stars within 21 light years can be reached in about 1.5 days of coasting. There is also time needed to decelerate the starship so the actual travel time would be longer. At 30,396c any point in the galaxy could be reached in about 3 years. *Thus Milky Way travel times become comparable to 16th century oceanic travel times via ships to various parts of the world!*

10.5.4 The Velocity Near the Singularity

In calculating the speeds at various times close to the singularity[38] at the 30 day point we have used an approximation to eq. 10.18. Letting $t = t_1 + \tau$ where τ is small, and letting $\Delta t = t_1 - t_0$ then eq. 10.18 becomes

$$
\begin{aligned}
v &= c\{1 - 2/(1 + ((c + v_0)/(c - v_0))\exp[2g(\Delta t + \tau)/(mc)])\} \\
&= c\{1 - 2/(1 - \exp[2g\tau/(mc)])\} \\
&\simeq c\{1 - 2/(1 - (1 + 2g\tau/(mc)))\} \\
&\simeq c\{1 + (mc^2/g)(1/\tau)\} \\
&\simeq c\{1 + (g^*mc^2/|g|^2)(1/\tau)\}
\end{aligned}
\qquad (10.32)
$$

Given the signs of g_1 and g_2 (eqs. 10.24 and 10.25) we see that

---

[38] The rapid increase in speed near the 30 day point takes place in "earth" time. On the starship the time contraction effect causes starship time to increase rapidly as pointed out in the discussion of acceleration in subsection 10.5.4.1. Thus the apparent increase in speed on the starship is much less rapid.

- For small negative τ both the real and imaginary parts of v approach $+\infty$ as $\tau \to 0$.
- For small positive τ both the real and imaginary parts of v approach $-\infty$ as $\tau \to 0$ from above.

as displayed in Figs. 10.6 and 10.7. A starship can decide to switch off engines and "coast" at high speed towards the destination at some time close to the singularity point.

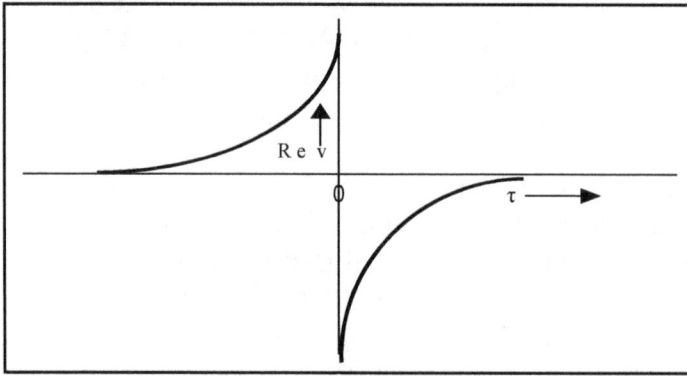

Figure 10.6. Qualitative plot of Re v from eq. 10.32 around the singularity at τ =0.

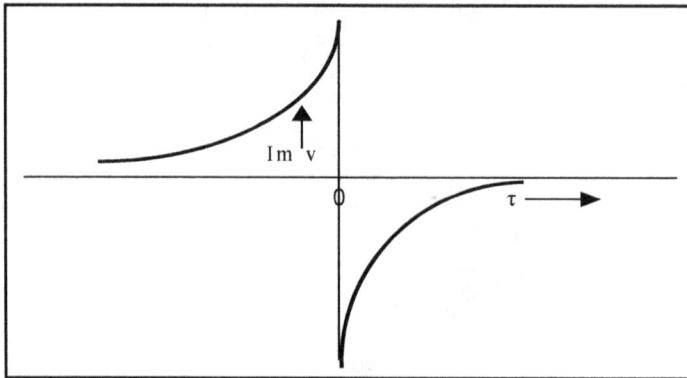

Figure 10.7. Qualitative plot of Im v from eq. 10.32 around the singularity at τ =0.

10.5.4.1 Acceleration Experienced on the Starship

The rapid acceleration, particularly in the neighborhood of $\tau = 0$ raises the question of the inertial forces that would be experienced by passengers on the starship. *Remarkably the acceleration that the starship occupants feel is quite moderate due to the effects of relativity.*

The calculation of the maximum acceleration begins with the inverse of the eq. 10.12 relativistic transformation from earth coordinates to starship coordinates:

$$F'^0 = \gamma(F^0 - \beta F^x/c) \qquad (10.12')$$
$$F'^x = \gamma(F^x - \beta c F^0)$$
$$F'^y = F'^z = 0$$

which implies the acceleration felt by occupants in the starship is

$$a' = F'^x/m = \gamma(a - v\gamma/c^2) \qquad (10.33)$$

where a (in the earth coordinate frame) is given by the derivative of eq. 10.18

$$a = dv/dt$$
$$= -4(g/m)((c + v_0)/(c - v_0))\exp[2g(t - t_0)/(mc)]/\{1 + ((c + v_0)/(c - v_0))\exp[2g(t - t_0)/(mc)]\}^2 \qquad (10.34)$$

$$\simeq 2c/\tau + 4(g/m)$$

At $\tau = 29.9999998$ days we find using the approach of eq. 10.32[39]

$$a = -2,891,516,763,121 + 72,672i$$

In ratio with the earth's gravitational acceleration g_E of about 980 cm/sec^2 we find

$$a/g_E = -2950527309+74i$$

---

[39] Kindly note that round off errors might lead the reader to slightly different results.

The large value of the real acceleration <u>in earth coordinates</u> might lead a reader to question the ability of starship occupants to survive. However if we transform the earth coordinates acceleration to the equivalent starship coordinates acceleration we find the acceleration felt on the starship

$$a' = a(5066^2c^2 - 1)^{-\frac{1}{2}} \approx a/(5066c) \qquad (10.36)$$
$$= (-1.94 \times 10^{-5} + i4.88 \times 10^{-13})g_E$$

has small real and imaginary parts compared to earth's gravitational acceleration. (We use the real speed in the example at τ = 29.9999998 days: 5066c approximately.) Thus the starship occupants will be safe.

We have developed starship dynamics to allow speeds far in excess of the speed of light for human occupants based on the ability to produce complex valued thrust and the ability to have complex velocities and space. We address these issues later in this chapter.

10.5.4.2 Travel Time Experienced on the Starship – Suspended Animation

Another issue is the travel time experienced by the starship occupants. By eq. 8.13 it will appear much, much longer than that measured on earth. For example if v = 5,000c then it will be 5000 times longer. A 2 month trip from earth's view would be around 1,000 years from the view of the occupants of the starship. *Therefore a practical method of suspended animation must be found. A 4 month round trip to a star would require the starship occupants to be in suspended animation for approximately 2,000 years – starship time. With suspended animation they would be biologically "in sync" with the earth measured travel time of 4 months (plus time spent at the destination) despite the starship elapsed time of 2,000+ years.* (See Appendix D for more detail.)

10.5.5 Constant Superluminal Starship Travel

Having reached an enormous *real* speed such as a speed between 5000c and 30,000c we can turn off the superluminal engines. If we didn't, then the real speed would rapidly decline after passing through the singularity generated when the thrust is complex.

The starship then moves at this constant speed in the absence of forces (and neglecting gravity and other minor perturbative forces). At a real speed of 5000c any place in the galaxy is a short travel time away.

And nearby galaxies are reachable as well. Figure 10.8 shows the time required to reach various interesting destinations at 30,000c.

Destination	Distance (ly)	Approximate Travel Time (years)
To the other end of the Milky Way Galaxy	100,000	3
To the Center of the Milky Way	30,000	1
Large Magellenic Galaxy	150,000	5
Small Magellenic Galaxy	200,000	7
Andromeda Galaxy	2,000,000	70

Figure 10.8. "Coasting" part of travel time to various destinations at a real velocity of 30,000c.

Since much, much higher "coasting" velocities are also possible, almost the entire visible universe becomes accessible to Mankind. Mankind has an incredible future if it has the will to seize it.

Starship travel will be in complex space. That could be an advantage. In real space, with which we are familiar, there are asteroids, planets, stars, nebulae, black holes, and dust. If we travel in real space in a straight line we will have a significant probability of colliding with some of these objects. We certainly would collide with interstellar and intergalactic dust. The dust alone could severely damage a starship traveling at high speed. Over long distances it would also reduce the speed of the starship significantly.

In complex space it is not known if there are any objects similar to those found in our real space. If it is empty then the starship will avoid all the pitfalls of travel in real space. Thus the starship speed will not be impeded and it will not be damaged or destroyed by collisions with matter. We can hope that this is the case.[40]

---

[40] This hope may not be realized. As pointed out in Blaha (2004) p. 78-9 matter might exist in complex space. The "Great Attractor", a seemingly empty region of space has recently been shown to be drawing an enormous number of galaxies towards it. The Great Attractor may be an extraordinary large mass located in complex space – not the real space of our experience. As pointed out in Blaha (2008), and his earlier books, it appears that the universe began as a complex space that, due to symmetry breaking, broke into disjoint real and complex parts with no interaction between them except gravity, and except in localized regions such as inside hadrons in which quarks and gluons enjoy a small region of complex space.

10.5.6 Deceleration a Tachyonic Starship to Sublight Speeds

Eventually all journeys end so we will now examine the deceleration of a starship as it approaches its destination. We turn on the superluminal engine. The thrust is reversed (g → –g) to decelerate. Otherwise the parameters of subsection 10.5.3 are in effect except v_0 which is now that of the 29.9999998 day point. We assume the starship was not slowed down by dust etc. and so traveled at constant velocity until the point when the superluminal engines were turned on to decelerate. The results are in the following table.

Example of Starship Deceleration

Time (days)	0	5	10	15	20	25	30
Re v	5067c	0.000249c	0.0001.33c	0.0001c	0.0000899c	0.0000893c	0.0001c
Im v	79,577,538c	3.73c	1.73c	1.00c	0.577c	0.268c	0.000001c

Note Im v in the table at the 30 day point is non-zero due to round off error. Im v is in fact zero at the 30 day point.

Note also that the speed is now real after deceleration, and has the original speed it had prior to acceleration.

10.5.7 Imaginary Space Travel Distance

The imaginary distance traveled during the trip is also an issue. The imaginary distance traveled must exactly cancel by the end of the trip so that the starship ends the journey in purely real space. Since the bulk of the distance traveled occurs during the "coasting" phase a starship cannot truly coast during the entire coasting phase. It must generate a thrust during the last part of the coasting phase with a value for g that brings the imaginary distance traveled to zero at the destination. During this interval the complex number g will have a negative value for g_2, and g_1 will equal zero.

Starting from eqs. 10.18 and 10.19 and assuming a purely imaginary thrust:

$$g' = ig_2 = in\pi mc/(2\Delta t) \tag{10.37}$$

where n is a positive, odd integer we can decelerate *after* nearing the real distance of the destination (approximately). To that end we need a large negative imaginary velocity that will undo the positive imaginary distance traveled during the "coasting" phase. And we must "coast" an imaginary distance to bring the cumulative imaginary distance to zero at the destination. The velocity equation (eq. 10.18) gives a large negative imaginary velocity, which can be seen by using a variation on the approximation of eq. 10.32:

$$v = c\{1 - 2/(1 + ((c + v_c)/(c - v_0))\exp[2g'(\triangle t + \tau)/(mc)])\}$$
$$\simeq c\{1 - 2/[(2c/(c - v_c)) + 2ig'\tau/(mc)]\} \qquad (10.38)$$

to order τ (for small τ) where v_c is the velocity during the coasting phase. Then since $|v_c| \gg c$ we can further approximate v to

$$v \simeq imc/(2g_2\tau) \qquad (10.39)$$

giving a large negative imaginary speed as τ approaches zero (the singularity point) from below.

If we now let the starship coast to a zero value for imaginary distance and follow the procedure of subsection 10.56 to obtain a zero value for imaginary speed we achieve our goal of traveling a very large real distance in light years in a relatively short time.

10.5.8 Starship Travel Summation

The preceding subsections have considered the case of a specific travel time of 30 days to near a singularity in v that enables us to achieve enormous superluminal speeds. We have also seen how to coast at these speeds and how to decelerate to small, sublight speeds. Other cases with different travel times to the singularity in v have similar qualitative features although the numerics are different. Planning a starship flight requires a complex program specifying a combination of all of these features of acceleration, coasting, and deceleration in such a way as to minimize energy and fuel usage.

Thus we have demonstrated that realistic superluminal travel is possible in principle if

1. Complex thrust engines are possible.
2. It is possible to travel in complex space.
3. Complex space does not have insurmountable obstacles to superluminal speeds such as interstellar dust or massive

objects that could lead to collisions or rapid erosion of the starship.

We will address item 1 in the remainder of the chapter and refer readers to Blaha (2008) and Blaha (2004) that suggests very strongly that complex space exists.

10.6 Complex Velocities for Quarks and Strong Interaction Gluons

The preceding sections of this chapter have assumed a complex starship thrust and the ability of the starship to penetrate complex space from the real space of our experience. Both of these assumptions are open questions at the moment. However the successful development of a derivation of the form of the Standard Model of Elementary Particles[41] based on a complex Lorentz group and complex space (but real time), and the simplicity with which the parity violating ElectroWeak sector emerges as well as the Strong Interactions, suggest that the ultimate basis of our universe is complex space. A complex space makes the distinction between, and the differing features of, the quark and leptonic sectors understandable in a remarkably simple way – usually the hallmark of a successful physics theory.

In Blaha (2008) we showed how to create second quantized tachyons (faster-than-light particles) using an enlarged Lorentz group and complex coordinates. Then we showed how the Standard Model of Elementary Particles emerged in a simple, natural way which, in itself, strongly supported our derivation. Nature favors simplicity – but in subtle ways.

As part of this derivation it became apparent that quarks and gluons had complex space coordinates and dynamics. (Leptons and the ElectroWeak bosons have real spatial coordinates with zero as the value of their imaginary coordinates.) The complex spatial coordinates of quarks led to a global SU(3) invariance, which in turn led to the Yang-Mills, non-Abelian bosons (gluons) that constitute the Strong interaction.

Another fact that emerged from the derivation was that "down-type" quarks were tachyons. Because free, strongly interacting tachyons

---

[41] Blaha (2008). See also Blaha (2004) for the General Theory of Relativity for a complex universe.

do not have a Lorentz invariant, filled Dirac sea like normal fermions, quark confinement to localized regions becomes a necessity. If tachyon quarks are bound into normal particles (hadrons) they are "stabilized" within the hadron. Thus our derivation led to the necessity of quark confinement to finite spatial regions, and complex momenta were thus also "confined" to quarks in localized spatial regions.

10.7 Quark-Gluon Plasmas

However if large atomic nuclei (such as gold nuclei) which contain many hadrons (neutrons and protons) collide at extremely high energy then an interaction region is created containing quarks and gluons in the form of a plasma that behaves like a perfect fluid according to data from RHIC (Relativistic Heavy Ion Collider) for $\sqrt{s}$ = 200 GeV Au–Au collisions.[42]

The quarks and gluons in an Au–Au interaction region must have complex spatial coordinates and momenta according to Blaha (2008). In current accelerators, and in accelerators that will come on line in the next decade, the characteristics of regions containing quarks and gluons moving in complex space will be further explored. The development of compact accelerators using novel techniques such as high-intensity pulsed lasers could lead to a generation of compact accelerators that could play a role in starship propulsion.[43]

Thus the ingredients are present for creating regions with quarks having complex coordinates with their motion generated by compact accelerators. Unfortunately, the presently envisioned accelerators will create only minute regions with single collisions – not enough to power a starship.

The current situation of quark-gluon plasma experiments is reminiscent of the initial experiments on atomic fission in the 1930's when experiments by Otto Hahn and Fritz Strassmann were instrumental in the discovery of uranium fission. Based on their work Lise Meitner and Otto Frisch developed a theory of uranium fission in January 1939.

---

[42] G. Agakichiev *et al* {CERES Collaboration], Phys. Rev. Lett. **92**, 032301 (2004); J. Adams *et al* [STAR Collaboration] Phys. Rev. Lett. 95, 152301 (2005); S. S. Adler *et al* (PHENIX Collaboration) arXiv:nucl-ex/05070004 (2005); P. Romatschke and U. Romatschke, Phys. Rev. Lett. **99**, 172301 (2007); K. Dusling and D. Teaney, arXiv:nucl-th/07105932 (2007); and references therein.

[43] G. Mourou, J. Rafelski, and T. Tajima, CERN Courier **49**, 21 (2009).

The subsequent development of macroscopic nuclear fission chain reactions led the way to atomic energy.

Our starship concept requires a major advance of a similar sort although on a much greater scale: a starship Manhattan Project to create starships with complex thrust generated by siphoning macroscopic amounts of quarks with complex momenta from a macroscopic plasma acceleration region.

10.8 Macroscopic Quark-Gluon Plasmas

If quark plasmas can be enlarged to macroscopic sizes then it becomes possible to consider using the plasmas to drive starships with complex velocities that can exceed the speed of light.

A possible basic design concept would consist of a circulating, self-intersecting, ring of heavy ions such as U^{238} nuclei. They would be rather like today's circular accelerators such as RHIC but intersect at many points to create another ring composed of quark-gluon plasma. This ring would in turn be accelerated by superconducting magnets to extremely high velocity. A part of the cycling plasma would be siphoned by magnets to generate rocket thrust. (See Appendix B for more detail.)

Current problems facing quark plasma rings include ring creation (Can we create a quark plasma ring?), ring persistence (How long can the ring persist?); and ring stability (Can the ring be stabilized to rotate in such a way as to achieve high circular velocities?). The relativistic effect of time dilation can help a relativistic circulating ring persist longer. Ring stability would seem to require powerful superconducting magnets that are beyond the state of the art at present.

Nevertheless there is a strong possibility for these technical achievements if the world makes a decision to invest in the future.

10.9 Generating Complex Thrust

Current accelerators accelerate particles to high, *real* velocities. The question arises how can a quark-gluon accelerator generate a quark plasma with high *complex* velocities? The solution will require a significant technical R&D effort to develop superconducting electromagnetic field technology to drive quarks to high complex velocity. We can illustrate the concept by considering the Lorentz force equation for a charged particle, such as a quark, with charge q:

$$dp/dt = qe(E + v \times B/c) \qquad (10.40)$$

where **E** and **B** are the electric and magnetic fields. Since the quarks and gluons in Blaha's derivation of the Standard Model have complex momentum, and thus complex velocity, eq. 10.40 shows that the effect of electromagnetic forces is to change both the real and imaginary values of the quark velocities. Creating the complicated electromagnetic force configuration necessary to drive a starship as described previously will require an intricate magnet assembly. The discussion of this topic is both premature and beyond the scope of this book. However we can get an inkling of the dynamics involved by considering the special case of a quark in a static, uniform magnetic field:

$$dp/dt = qev \times B/c \qquad (10.41)$$

with the quark energy satisfying

$$dE/dt = 0 \qquad (10.42)$$

In view of the constancy of the energy in this example, the absolute value of the velocity, and γ, are constant although the direction and complex values of the velocity change. As a result eq. 10.41 has the solution:

$$v = v_3\varepsilon_3 + r_g\omega(\varepsilon_1 + i\varepsilon_2)\exp(-i\omega t) \qquad (10.43)$$

with a triad of real orthonormal unit vectors: ε_3 which is parallel to **B**, and ε_1 and ε_2 which are orthogonal to **B**. The constant ω is the precession frequency and r_g is the radius of gyration. The picture that emerges is a fixed radius counterclockwise spiral around the direction of **B**.

However the important result of this example from our point of view is the complexity of **v**. Electromagnetic fields can be used to accelerate and change the initial complex velocity of quarks. Thus we have a mechanism for complex starship thrust using electromagnetic acceleration of quarks in the circulating quark-gluon plasma rings.

10.10 A Simple Quark Drive Starship Design

The design of a starship is a difficult matter. Currently it is an impossible matter since there are so many engineering and scientific unknowns at present. However one can describe a likely overall design – if the scientific and engineering breakthroughs described above take place. We will propose such a simple overall design in this section.

1. A starship would have two separate relativistic heavy ion colliders that would accelerate and collide heavy ions such as U^{238} at high energy (much greater than $\sqrt{s} = 200$ GeV) at multiple points to produce 2 rings of quark-gluon plasma. Each colliding heavy ion ring would look like a braided ring. (See Fig. B.1.) Initially each quark plasma ring would consist of discrete plasma points. Acceleration of a quark plasma ring would cause the ring to become continuous through smearing effects.

2. Each heavy ion ring/quark ring pair constitutes a *quark drive engine*. The quark plasma in the two engines will be traveling in opposite directions. Quarks in each quark ring would be accelerated to a high complex velocity comparable to the values in eq. 10.31.

3. We decided on a tangential siphoning of a ring since it would require less powerful magnets and less energy. Each ring would be siphoned at points that are diametrically opposite and thus both points of siphoning on each quark ring would contribute to the starship's thrust although in opposite directions. See Figs. 10.9 and B.2. Two siphoned thrust points (one from each quark plasma ring) would move the starship in one direction. Thrust from the other two siphon points would move the starship in the opposite direction.

4. A superluminal quark drive would have two quark drive engines having a total of four rings.

5. Naturally it would be reasonable to have the rings surround the outer edge of the starship to reduce the required bending magnet strength requirements.

A

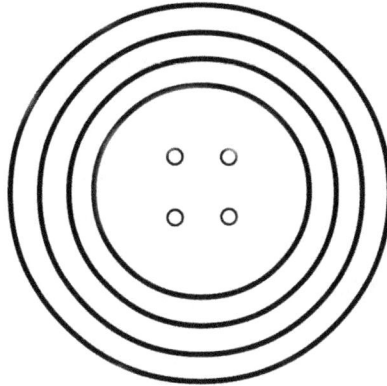

B

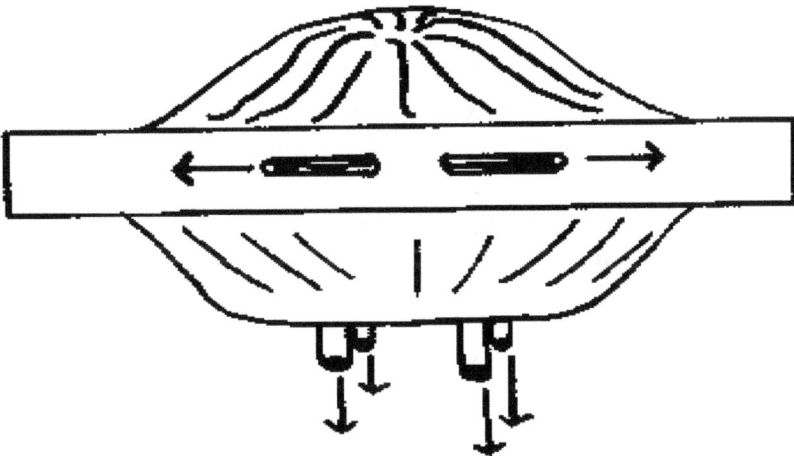

Figure 10.9. Views of a simple design for a starship using quark drive for interstellar travel and nuclear (or fusion) rockets for local (interplanetary) travel. A. Simplified schematic of the 4 plasma rings with tangential thrust points of quark plasma rings indicated by arrows, and 4 nuclear engines in the center. B. Side view of Starship showing direction of thrust from the quark drive (from the enclosing quark plasma rings) and nuclear engines (under the ship). The central portions of the starship contain the items listed in number 6 (next page). Drawings by Mr. Robert J. Hutchinson.
© S. Blaha (2009). See Figs. B.1 and B.2 for a more detailed view of the braided rings.

6. The interior region of the starship would hold fuel, life support equipment and cargo, nuclear/fission power generators, living quarters, cargo areas, and (perhaps four) nuclear engines for "short distance" travel within star systems.

The result is a saucer-shaped starship design depicted in Fig. 10.9 and on the front cover of the book. The author would like to point out that the geometry of this design is motivated by physical principles[44] and *not* by UFO depictions.

10.10.1 The Starship's Nuclear Rockets

Nuclear rockets would be used for "short" distance travel within solar systems. These rockets could be either ion drive rockets where ions were accelerated by electromagnetic forces generated by electricity provided by the ship's nuclear reactors.

Or the rockets could simply use nuclear reactors to directly heat a fuel such as hydrogen whose ejection would provide thrust for the starship. The decision will be based on maximizing the efficiency of the nuclear rockets.

The starship is not intended to land on large planets. It would contain shuttles for that purpose.

---

[44] For example the disc-like shape of the starship reflects the probable large width (perhaps as much as, or more than, 20 km in diameter) of the accelerator rings, which would be situated along the circular edge of the disc. The larger the rings the lower the requirements on the magnets that guide the particles around the rings. A spherically shaped starship of that size appears to be wasteful. So a disc-like shape with a central bulbous region for the crew, fuel, and cargo seems more practical and economical.

11. Exploration Program for the Stars and Galaxies

11.1 How Do We Start?

An interplanetary exploration and colonization program was outlined in earlier chapters. This ambitious program will undoubtedly take many years to complete.

While that program is being implemented we need not wait and defer a program to research, design and build a starship along the lines described in the preceding chapter. We can define a multiphase series of steps that should be taken that will ultimately lead to a quark drive starship within a reasonable time frame. Of course, the starship program assumes the correctness of the conditions stated in section 10.5.8, which we restate here for the reader's convenience:

1. Complex thrust engines are possible.
2. It is possible to travel in complex space.
3. Complex space does not have insurmountable obstacles to superluminal speeds such as interstellar dust or massive objects that could lead to collisions or rapid erosion of the starship.

The Starship Project is rightly an international effort that will be spearheaded by the leading nations. It will be very expensive but the payback for a successful effort will be an Open Door to the Universe. Perhaps this is the best hope for the future of Mankind.

11.2 Planet Search Phase

The search for extraterrestrial planets is currently underway. New observation instruments and satellite observatories are under development that will enable us to find earth-like planets near stars that might be capable of supporting human life.

There are about 100 stars within a 21 light year radius from earth. The proposed starship with its high superluminal speed capabilities will enable trips to stars within 1,000 light years of earth with short travel times.

Since the research, design and development of a starship will probably take decades of effort, astronomers will have time to develop a priority list of stars to visit that could provide homes for Mankind. The starship could then perhaps visit and explore the top ten stars on the priority list. If truly earth-like planets are found then the construction of a fleet of starships and the creation of initial colonization outposts can then begin.

11.3 Quark Plasma Research Phase

The RHIC, which has successfully created interaction regions containing quark-gluon plasmas, should be the first of a series of increasingly powerful, circular accelerators that ultimately will produce a macroscopic, although still small, regions containing quark-gluon plasmas that can be studied to determine if the Standard Model theory of Blaha (2008) was correct in predicting quarks (and gluons) have complex spatial momenta.

These accelerators should also be built with the goal of studying improved acceleration techniques (and superconducting magnets) that produce high velocities in accelerators that are physically small in radius and also require less power consumption to operate.

When acceptable accelerators are built then a "braided accelerator" (See appendix B.) should be developed that has multiple interaction points that are so dense that a ring of quark-gluon plasma is produced when magnets start the dense ring of isolated regions of quark-gluon plasma to rotate (accelerate). Smearing effects should produce an effectively continuous quark-gluon plasma ring. Research efforts should be directed to efficient acceleration of the quark-gluon plasma ring using magnets to accelerate the quarks (which are charged), which in turn can drag gluons along with them (due to the Strong Interaction) as the quarks accelerate. Designing the accelerating quark-gluon plasma rings to be stable for "long" intervals is also an important research goal.

The successful creation of extremely high velocity, stable, rotating quark-plasma rings, which are macroscopic in size and density, and also energy efficient from a starship design perspective, would enable us to develop superluminal engines for a starship.

11.4 Design and Development of Plasma Ring Thrusters Phase (Quark Drive)

The successful development of a stable, high energy, high quark density quark-gluon ring is the key to quark drive. The starship thrust must be complex in nature to achieve the high superluminal speeds (as seen in detail in the preceding chapter). A quark-gluon ring of the type described above can have a fraction of the circulating quarks siphoned off tangentially (to save energy) to furnish a complex thrust from the complex momentum of the siphoned quarks.

Since we do not want the thrust to cause the starship to rotate, we need two quark-gluon rings with quarks circulating in opposite directions to furnish thrust from diametrically opposite sides of the rings. (Fig. 10.9) To give greater mobility to the starship it makes sense to have siphoning at diametrically opposite points on each quark-gluon ring. Thus one has "forward" and "reverse" capabilities by selecting the appropriate siphon on each ring.

The fuel (perhaps U^{238} or some other heavy ion), magnet support equipment, nuclear (or fusion) reactors to generate power for the rings, crew quarters and supplies as well as cargo space can be situated within the set of rings. (Fig. 10.9)

11.5 Nuclear or Fusion Engine Phase

In addition nuclear or fusion engines are required for local movement within a star system. Presumably these will be developed as part of the interplanetary exploration program described in earlier chapters. It would seem reasonable to have four nuclear (fusion) engines arranged in a cluster to provide redundancy for reasons of safety in case one or more engines failed. The exact design of these engines remains to be determined. Preliminary designs do exist from NASA work in this area in the 1950's and 1960's.

11.6 Long Term Suspended Animation

As described in section 10.5.4.2 biomedical engineering must develop a means of long term suspended animation with a minimal initial goal of 1,000 years. The starship occupants would be put in suspended animation during their trip to counteract the effects of time dilation (eq. 8.13).

They would sleep for many years during the trip so that upon their return to earth their biological clocks would be in sync with the elapsed earth time although the elapsed starship time could be many years. See section 10.5.4.2 for an example. (See Appendix D for more detail.)

11.7 Starship Prototype and Test Phase

After the R & D phase a "small" prototype for test purposes should be constructed and tested within the outer solar system. Study of test results often leads to improvements both in performance and safety.

11.8 A Working Starship Ready to Explore

After successful prototype tests a full scale starship should be built – probably in orbit around the moon. This ensures that any accidents (particularly nuclear accidents) will not impact on the earth.

And then its off to the stars!

11.9 Concluding Remarks

This book outlines an ambitious program to conquer the solar system and the universe that ultimately includes a massive effort to transfer people from earth to other earth-like planets just as a massive transfer of people from Europe led to the growth of the Americas. We began with a consideration of the declining state of the earth under the crushing burden of a large population and increasing social demands that, if pursued, would bring the progress of civilization to a standstill creating a static civilization – equally involved in production and waste disposal. Surely, Mankind should strive for a better fate.

The only serious long-term option for Mankind is to make a major space effort both in the Solar System, then the galaxy, and then the universe. So the universe will *not* be "wasted" space.

The price is high – perhaps half a trillion dollars – but doable. The price for not acting is much higher.

Appendix A. Preliminary Tests of Concept and Project Cost

It is difficult to estimate the R&D costs of our proposed starship. Clearly it will be large and a guesstimate of half a trillion dollars is reasonable. However, this sum will be spread over a fairly long time period of thirty or more years because of the magnitude of the research and development effort needed. And the project will have critical points where success is required to pass on to the next stage.

At the time of this writing the United States appears headed towards a total deficit of ten or eleven trillion dollars over the next ten years. If the cost for starship R&D is of the order of half a trillion dollars (a small amount relative to the projected US deficit considering the reward for success), and spread over thirty years, then the average cost on a yearly basis would be slightly more than seventeen billion dollars per year. The initial yearly expenditures will tend to be much smaller. The expenditures in the latter stages of the project will be much larger. The cost of the starship program is comparable to NASA's current budget of about eighteen billion dollars a year. It is affordable for the United States. For political and economic reasons the formation of an international consortium of the large space-faring nations to fund and execute the project would significantly reduce the per country costs. If the United States paid sixty per cent of the costs then its share would be about ten billion dollars per year. Needless to say, the project would also serve as a major unifying force for humanity.

Before plunging into massive spending there are three fundamental requirements that must be met. Fortunately, these requirements will not require massive initial expenditures – only a few billion dollars to the test for complex quark momenta. (See section A.1 below). Much of it has already been spent for current accelerators and the LHC at CERN which is about to begin operation. A reanalysis of deep inelastic electron-nucleon scattering might also help determine if quarks have complex momenta. If the test are successful then the path to starship development receives a major boost.

The three fundamental requirements for the new type of starship that we have proposed are:

1. Complex thrust engines (based on complex quark momenta) are possible.
2. It is possible to travel in complex space.
3. Complex space does not have insurmountable obstacles to superluminal speeds such as interstellar dust or massive objects that could lead to collisions or rapid erosion of the starship.

A.1 Complex Quark Momenta Experiments

The first requirement can be tested at existing accelerators and at CERN's new LHC particle accelerator scheduled to become operational in the fall of 2009. It can also be tested by a reanalysis of spin-dependent deep inelastic electron-nucleon scattering data acquired at SLAC (the Stanford Linear Accelerator) and other accelerators. The first result that must be proven is that quarks have complex spatial momenta. (Complex spatial momenta means that the x, y, and z components of a particle's momentum are each complex numbers – not the real numbers with which we are familiar in ordinary life.) Since quarks are confined within hadrons such as protons and neutrons their momenta cannot be directly measured. Furthermore, since the real and imaginary parts of a particle's complex momentum are separately conserved the real parts of a quark's momentum will appear to be completely normal and the physical law of conservation of total momentum will be satisfied for the total real momentum. The total imaginary momentum will also be conserved. So the parton model studies of deep inelastic electron-proton scattering will continue to be correct even if quark's also have an imaginary momentum as well. However the spin dynamics of quarks differs if quarks have complex momenta. The spin s_μ and momentum p^μ of a spin ½ particle such as a quark satisfy the relation

$$s_\mu p^\mu = 0 \qquad\qquad (A.1)$$

If the momentum is complex then the spin is also complex in general. Consequently, the equation A.1 interrelates the real and imaginary parts of a quark's spin. One of the long standing problems of the analysis of the spin of quarks within hadrons in deep inelastic scattering has been the whereabouts of the spin. Parton studies of spin-dependent scattering have shown inexplicable features which have led to questions such as "Where is the spin hiding?" We suggest the imaginary part of quark spin, which has not hitherto been part of theoretical studies, may account for the "nucleon spin puzzle."[45]

New LHC results may also confirm that quarks have complex spatial momenta. If no evidence for complex spatial momenta is found then there would be no point in proceeding with the proposed starship program. The success of Blaha's derivation[46] of the Standard Model strongly suggests that quarks will be found to have complex momenta.

If quarks have complex spatial momenta then it will be possible to build quark drives that produce complex thrust and go past the speed of light barrier. Complex thrust engines become feasible.

A.2 Traveling in Complex Space

Quarks are confined within hadrons as far as we know. It is thought that quark confinement is due to the strength of the Strong Interaction – if we try to knock a quark out of a proton or some other hadron then a quark – antiquark pair will pop out of the vacuum as the quark separates from the rest of the hadron. As a result the quark that was knocked out of the hadron is paired with one of the created pair of quarks with the result that an isolated quark cannot be created.

Thus if quarks have complex momenta the regions of space containing particles with complex momenta are very small – the hadrons and the interaction regions when hadrons undergo collisions. If hadrons contain particles with complex momenta they must also be regions of space with complex coordinates – islands of complex coordinates within the strictly real space with which we are familiar. These islands are small presumably due to the Strong Interaction as noted before. If we can create a very large (macroscopic) region through the combined interaction of many hadrons then we have a macroscopic region with

---

[45] See M. Alberg, J. Ellis and D. Kharzeev, CERN-TH/95-47 (1995) and references therein. There is an extensive literature on the nucleon spin puzzle up to the present day.
[46] See Blaha (2008).

complex coordinates. So it would seem likely that the universe is a complex space – but we are unable to enter it because all familiar matter is in the real part of the space and there is no way for ordinary matter to collide or otherwise enter the non-real part of the space. We can think of our real space as a surface within complex space, to which we are confined due to a lack of complex forces that would generate complex accelerations and complex momenta thereby.

Using a quark plasma with quarks of complex momentum to generate thrust (a complex force) would enable us to enter complex space.

A.3 Obstacles in Complex Space?

Several immediate questions arise about complex space. The first question is whether we can see things in complex space from our position in real space. The answer appears to be yes. Blaha (2008) develops the quantum field theory of a Standard Model with quarks with complex spatial momentum. These quarks can interact with photons (light) and thus can be "seen" within protons and other particles despite their complex momenta. Thus we can see matter in complex space.

However we can only measure the real part of its location and the real part of its momentum. Thus matter in complex space would appear to be in the real part of space.

If a starship were traveling through complex space as we discussed in earlier chapters then possibly matter might be encountered. But since quarks and gluons are confined to within hadrons, and since hadrons are very small, elementary particles in real space only, they would not be an obstacle for a starship. In fact, starships would benefit from traveling in complex space because they would avoid interstellar dust and large objects located in the real part of space. There is one caveat. If quark stars exist, and that remains to be proved, then the interior of the quark star would be in complex space and an encounter with a starship would be fatal to the starship. Thus there is a potential issue of obstacles in complex space. However, a quark star would be detectable from the vantage point of our real space. So we can be fairly certain that quark stars are uncommon or non-existent.

We conclude obstacles in complex space are a minimal (or non-existent) hazard for starships. And avoiding the matter in real space is a great advantage for starships.

Appendix B. Complex Thrust Using Braided Accelerators

In earlier chapters we described our proposal for starships based on a quark drive technology using circular accelerators to accelerate U^{238} (or other heavy ions) to extremely high energy, have collisions between the U^{238} ions at multiple points in the accelerator ring that generate a quark – gluon plasma that is then fed to a ring where the quark plasma is accelerated. Part of the accelerated quark plasma is siphoned off to provide thrust for the starship – we call it a *quark drive*.

In this appendix we will describe the concept of our quark drive in more detail. Figure B.1 shows one pair of accelerator rings – one quark drive engine. The central circular ring contains an accelerating quark-gluon plasma. Superconducting magnets (not shown) take the quark-gluon plasma generated by the many collision points of the "oscillating" ring of heavy ions such as U^{238} ions and cause the quarks to accelerate "dragging" gluons along with them due to the Strong interaction between quarks and gluons.

The oscillating (braided) heavy ion ring accelerates heavy ions and (as shown) has points of "self interaction" where the ions collide and create "droplets" of quark – gluon plasma[47] in the quark plasma ring which is undergoing rapid acceleration to speeds far beyond the speed of light. The heavy ion ring is, as its name suggests, one ring of heavy ions that intersects with itself as shown.[48] Its "oscillating" path requires powerful magnets. The superconducting magnets required for these processes must be significantly more powerful then the most powerful superconducting magnets now available.

---

[47] Since the quark gluon plasma transitions with extreme rapidity to a burst of normal particles the acceleration of the droplets and their confinement within the ring must be extremely rapid. This requirement will pose major R&D challenges.

[48] At each intersection point the quark plasma ring magnets will extract the quark-gluon plasma created at that point and inject it into the ring of circulating (and accelerating) quark plasma. The extraction time will have to be very short so the magnets performing the extraction will have to be very powerful.

At diagonally opposite points quarks can be siphoned off to give a plasma thrust with complex momentum. The siphoning points are tangential to the ring to save in the magnetic energy required.

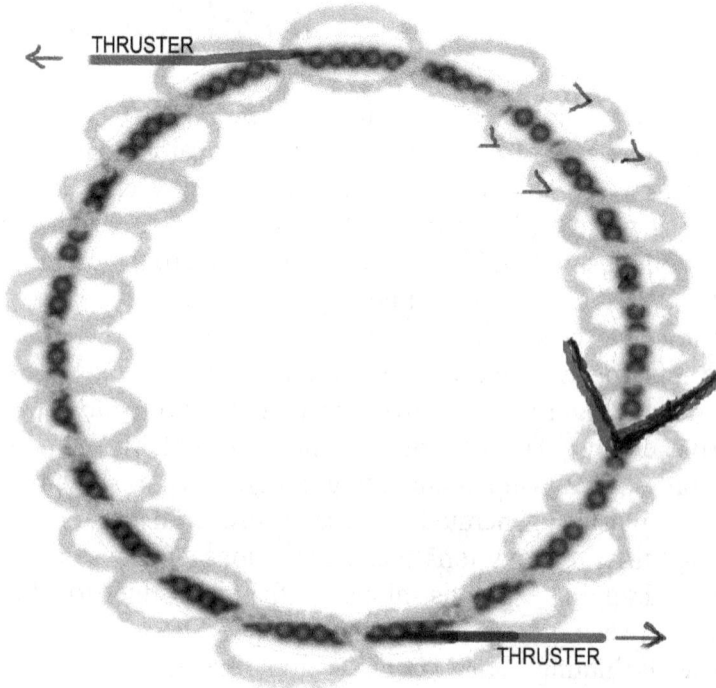

THRUSTER

THRUSTER

Figure B.1. A pair of accelerator rings that constitute one of the 2 quark drive engines for a starship. The large thick ring holds the accelerating quark plasma. The ring that oscillates around it accelerates heavy ions such as U^{238} ions. This ring self-collides at points along the quark plasma ring to feed quark-gluon plasma to the quark plasma acceleration ring. Superconducting magnets that accelerate and guide the flow of the rings are not shown. The thrust ports siphon quark plasma from the quark ring creating quark "ion" streams to power the starship. Either one or the other port siphons the quark ring depending on whether we want "forward" thrust or "rearward" thrust. There is another similar pair of rings (the other quark drive engine) that together with this engine provide forward or rearward thrust from diagonally opposite tangent points on the quark rings. Arrows indicate the direction of flow.

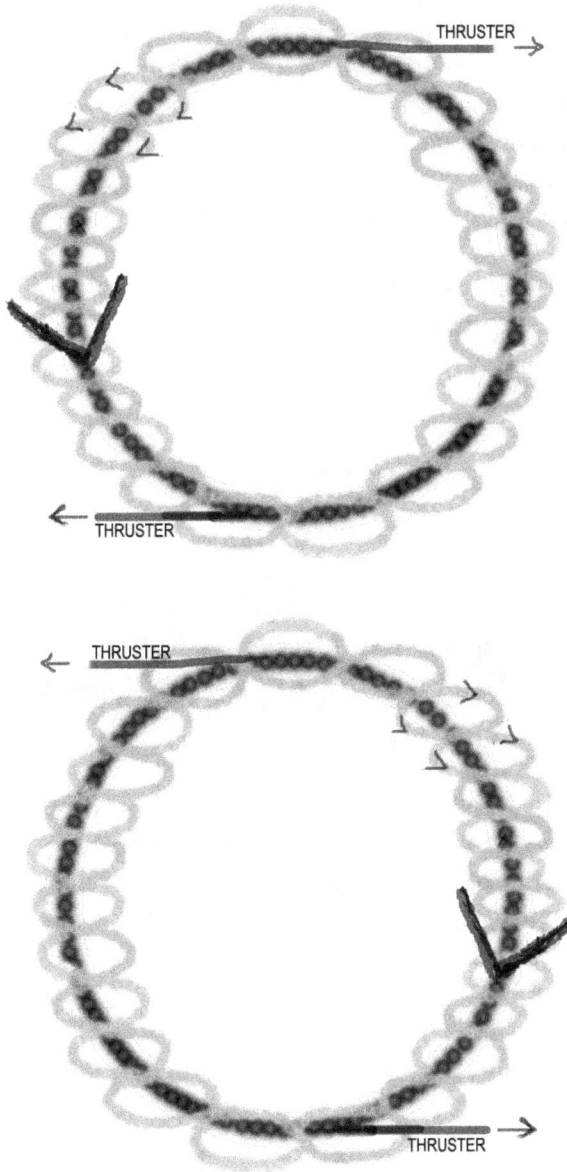

Figure B.2. The two quark drive engines that power a starship. One would be above the other on the starship rim to provide thrust from diagonally opposite tangential points in the "forward" or "rearward" direction.

As Fig. B.2 indicates we have two quark drive engines. They are stacked vertically as suggested in Fig. B.3. There are several reasons for having two engines. One is redundancy for safety purposes. If one engine fails the other engine will hopefully enable the ship to limp to a destination or repair station. The other reason for two engines is to save fuel for the nuclear or other orienting rockets that would have to "turn the starship around." Presumably the starship will be very large and turning it around and other maneuvers will require significant amounts of fuel.

Thus we have good reason for our overall design. It will of course require an enormous R&D effort to make working quark drives.

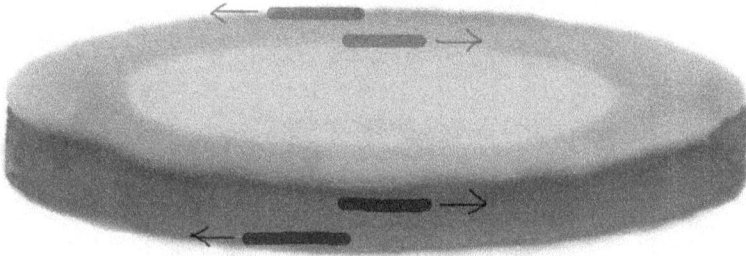

Figure B.3. The two engines are stacked as shown with centers on the same vertical line to provide a complete quark drive with both forward and backward capabilities. The arrows point in the direction of the thrust from each siphon point on the two engines.

Appendix C. Seeing and Navigating through the Cosmos on Superluminal Starships

The view of the universe that a starhip crew sees when the starship is traveling faster than the speed of light is very different from the view of a spaceship traveling at low speeds of a few tens of miles per second.

As Mallove (1989) points out[49] an observer on a starship traveling at a relativistic speed near but below the speed of light will see the visible stars and galaxies compressed to within a cone in the direction of the starship (Fig. C.1). The cone gets more narrow as the speed of light is approached due to aberration and in the limit as the speed approaches the speed of light becomes a point directly ahead of the starship.

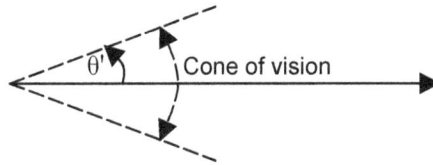

Figure C.1. Cone of vision around direction of starship motion in the starship coordinate system with the angle θ' determined by eq. C.1 for sublight starship speeds.

The relativistic equation for aberration is

$$\cos \theta' = (\cos \theta + \beta)/(1 + \beta \cos \theta) \qquad (C.1)$$

---

[49] pp. 182-185. They reference the scientific papers that are the basis of his description.

where θ is the angle of a star (or galaxy) relative to the starship's direction of motion as measured in the earth coordinate system and θ' is the angle of a star (or galaxy) relative to the starship's direction of motion as measured in the starship's coordinate system.

The inverse relation is

$$\cos \theta = (\cos \theta' - \beta)/(1 - \beta \cos \theta') \qquad (C.2)$$

C.1 Sublight Case: $\beta < 1$

As $\beta \rightarrow 1$ (the speed of light) eq. C.1 indicates $\theta' \rightarrow 0°$ showing the entire view of the universe is compressed to the exactly forward direction. Fig. C.1 shows the cone of visibility for a spaceship traveling near the speed of light at perhaps .6c - .9c. The cone angle θ' satisfies

$$\cos \theta' > \beta \qquad (C.3)$$

The rest of the field of view of the starship is total blackness except the point in the directly rearward direction ($\theta' = 180°$) for any object at $\theta = 180°$.

C.2 Superluminal Case: $\beta > 1$

For $\beta > 1$ eqs. C.1 and C.2 still hold and there is a cone of visibility similar to that depicted in Fig. C.1 . However the cone angle θ' for superluminal speeds, $\beta > 1$, satisfies the relation

$$\cos \theta' > 1/\beta \qquad (C.4)$$

The rest of the field of view of the starship is total blackness, as in the sub-light speed case, except the point in the directly rearward direction ($\theta' = 180°$). We note that as β gets very large the cone of visibility becomes larger. At $\beta = \infty$ the cone of visibility becomes the angular region between $\theta' = 0°$ and $\theta' = 90°$ (the forward hemisphere).

C.3 Superluminal Starship Visibility

As a result visual navigation at high superluminal speeds becomes difficult although one can conceive of electronic imaging that "undoes" the effects of aberration and enables visual navigation.

A further problem is the location of a destination. If we send a starship from the earth to a star, for example 30 light years away, we have to project the location of the star at the time the starship arrives based on the star's current motion as determined by earth observation. If the motion of the star is modified by the gravitation effects of other nearby stars during the 30 years that the light from the star was traveling to earth, or if the star's motion is not accurately determined, a starship could arrive at a point that is some distance from the star.

Thus navigation to a destination is a significant issue.

C.4 Effect of Doppler Shift at Superluminal Speeds

A starship traveling at relativistic sublight speeds will see stars having their color changed significantly due to the Doppler Shift effect. At superluminal speeds the Doppler Shift will also have a significant effect that will change the view of starship occupants. This issue is again surmountable if we use electronic imaging techniques to "undo" the Doppler shift and thus display stars as they normally look in the visible human frequency range.

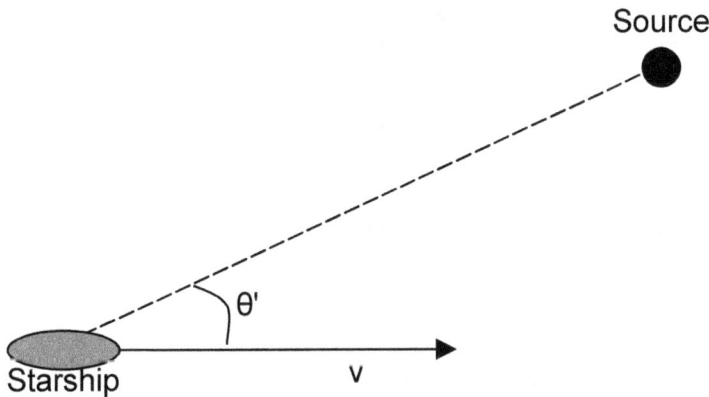

Figure C.2. The angle of the source θ' with respect to the starship's velocity **v**.

The relativistic Doppler shift for sublight speeds of a light wave of frequency v is given by

$$\nu = \nu_0(1 - \beta^2)^{\frac{1}{2}}/(1 - \beta \cos \theta') \qquad (C.5)$$

where ν_0 is the frequency of the light emitted by the source and θ' is the angle of the source relative to the starship's velocity (Fig. C.2). For superluminal speeds the Doppler shift is

$$\nu = \nu_0(\beta^2 - 1)^{\frac{1}{2}}/(\beta \cos \theta' - 1) \qquad (C.6)$$

This can be seen by considering an electromagnetic plan wave, which is a combination of

$$\cos[(k \cdot x - \nu t)/2\pi] \qquad \text{and} \qquad \sin[(k \cdot x - \nu t)/2\pi] \qquad (C.7)$$

Upon transforming from the earth coordinate system, for example, to a coordinate system moving in the x-direction at a speed faster than light (See eqs.) both the energy (ν' up to a constant) and the time t' obtain a factor of i (that cancel each other) so eq. C.6 is the correct frequency in the superluminal frame. The sign of the frequency is always positive by convention due to the form of electromagnetic waves and eq. C.4 dictates the form of the denominator in eq. C.6.

For large $\beta \gg 1$ eq. C.6 becomes approximately

$$\nu \approx \nu_0/\cos \theta' \qquad (C.8)$$

In the forward direction $\theta' = 0$ the Doppler shift goes to zero. Due to eq. C.4 the maximum value of the Doppler shift for large β in the field of vision is

$$\nu \approx \beta \nu_0 \qquad (C.9)$$

So the "wide" angle electromagnetic waves are shifted to large frequency.

Eq. C.6 and the discussion that follows suggests that frequency shifts will be substantial for extremely fast starships. The result will be a distorted view of the universe. However electronic imaging techniques can be implemented to restore the view that humans would normally see. Thus the combined effects of aberration and the Doppler shift on the view of space from the starship bridge can be electronically corrected to give a "normal" view of the oncoming space.

Appendix D. Suspended Animation

Due to time contraction on starships traveling much faster than the speed of light a passenger on a starship would age much more quickly than their counterparts on earth. If a starship were traveling at 5,000c (5,000 times the speed of light) time on the starship would be accelerated by a factor of 5,000 according to eq. 8.13. Thus a time span of one year on earth would correspond to a time span of 5,000 years on the starship. While it is possible to have a large multi-generation starship it is far more desirable to have a starship where the time intervals experienced by occupants is less than or equal to the corresponding earth time interval. This requirement would be met by a form of suspended animation that brings bodily processes to a halt during a trip with revival at the destination.

Then people taking a trip to a star that takes 2 earth years at, for example, 5,000c would not age at all during the trip. The time interval spent at the destination would equal the corresponding earth time interval. If the return to earth were in the same manner as the trip to the destination then the people taking the trip would have aged biologically approximately the same amount as earth people.

Suspended animation has been a subject of investigation for some time. Most work has focussed on slowing down life processes in small mammals by injecting them with chemicals and lowering their body temperature in a controlled manner. Ideally, it would be best if a "miracle" antifreeze were found that, when injected into humans in low concentration, would allow body temperature to be lowered to the point where all bodily functions stopped without any damage to the body. Revival would then amount to warming the bodies, perhaps "shocking" the heart into beating, and then injecting the revived people with a chemical that neutralized or eliminated the antifreeze. A minimal requirement of the antifreeze would be to "prevent" bodily fluids, bones, and so on from expanding and cells from rupturing.

Initially, when travel would be to nearby stars, the period of suspended animation would only need to last for a few decades at most.

In the case of travel to far stars in the galaxy or to other galaxies the period of suspended animation could be up to millions of years. The creation of suspended animation chambers to hold people at low temperatures for long periods of time and to revive them at the destination is a major technological challenge that will probably require the development of new materials. Interestingly, new boron-based materials that are harder than diamonds have recently been found. New metal alloys of great durability can also be developed. As long as the starship hull is not breached maintaining a low temperature should not be too difficult.

The question of reviving a person of suspended animation is a difficult one that will require a significant research effort. At first glance it appears that the chamber a person in suspended animation occupies will first be filled with pure oxygen, and then the heart and lungs will be started followed by the other organs (if they do not start automatically).

Appendix E. Engineering Very Long Life Starship Components

At high speeds in excess of $\sqrt{2}c$ (approximately 1.414 times the speed of light) time on a starship proceeds at a speed much faster than time does on earth. For speeds very much greater than c starship time $t' \approx \beta t$ where $\beta = v/c$ and v is the speed of the starship. Thus a starship traveling at 5000c has its time increasing at a rate 5,000 times the rate of time increase on earth.

If we travel the approximately four light years to the nearest star at a speed of 4c then it will take approximately one year to reach the star in earth time but the starship occupants will experience approximately four years of travel and thus age four years. The components and equipment on the starship will also age four years. For this short trip at "low" speed the time expended on the starship is not excessive.

But suppose we consider a trip to a far part of the galaxy 20,000 light years away at a speed of 5,000c. Then the earth time expended will be four years but the starship time expended will be $4 \times 5,000 = 20,000$ years. Trips to other galaxies would require millions of years of starship time although the earth time expended would be small. For example, to visit the Andromeda Galaxy 2,000,000 light years away at a speed of 1,000,000c would require two earth years and 2,000,000 starship years.

As a result for trips to distant stars and galaxies the components and equipment of the starship must retain their integrity for up to millions of years. (We are neglecting the possibility that dust in the complex part of the cosmos would erode the starship significantly.)

At present we know that some materials can last millions of years and retain their form albeit in a chemically changed form. Fossils from many millions of years ago prove it. The challenge is to develop materials and instrumentation for starships that retain their shape and functionality for exceedingly long times. As mentioned in the prior appendix new super strong materials are being created in the laboratory. A major effort to make the required type of materials will undoubtedly prove successful.

Appendix F. Robot Guidance and Robot Exploratory Starships

While the occupants of a starship sleep in suspended animation sensors will be required if matter exists in complex space and also to monitor the starship's motion. Computer programs will be needed to guide the starship's motion turning the starship's engines on and off, and regulating the thrust. Again we see that current computer technology would probably suffice for short trips to the stars. But extremely long life stable computers would be required for travel to far stars and galaxies. Creating computers with working lifetimes of up to millions of years is a major research challenge. Since the computers will be in a very low temperature environment the challenge of building long life computers is somewhat reduced.

Prior to manned starships it seems reasonable to send computer-guided, robot starships on round trips to nearby stars to explore as well as test the starship. This period will be similar in spirit to the current probes being sent to Mars and the outer planets.

Appendix G. Fuel Consumption on Starships

The high speeds attainable through the quark drive mechanism naturally raise the question of the amount of fuel required to reach those speeds. A major advantage is the high speed of the quarks ejected to produce the starship thrust.

We shall use the equations describing superluminal starship dynamics starting from eq. 10.10 and consider them for the numerical example in section 10.5.3. In this example, the thrust was generated by ejecting a constant amount of mass per second at high velocity:

$$dm/dt = 1 \text{ gm/sec} \qquad (10.29)$$

We can calculate the total mass ejected to obtain a specific starship speed from

$$\Delta m = \int_{t_0}^{t'} dt' \, dm/dt'$$

$$= (t' - t'_0) \times 1 \text{ gm/sec} \qquad (G.1)$$

From eq. 8.13 we can relate starship time intervals to earth time intervals so that eq G.1 becomes

$$\Delta m = (\beta^2 - 1)^{\frac{1}{2}} (t - t_0) \times 1 \text{ gm/sec} \qquad (G.2)$$

If t t_0 is approximately one month and $\beta = 5{,}000$ then

$$\Delta m = 6480 \text{ metric tons} \qquad (G.3)$$

for a 100,000 metric ton starship. If $t - t_0$ is approximately one month and $\beta = 30,000$ then

$$\Delta m = 32,400 \text{ metric tons} \qquad (G.4)$$

for a 100,000 metric ton starship. Thus the ejected mass in both cases is a respectfully small portion of the total starship mass – especially in comparison to present day chemical rockets.

We conclude that thrust matter consumption is favorable for quark drive starships.

G.1 Starship Energy Source

At this point in time the only feasible energy source for starships is nuclear energy. It is reasonable to expect that fusion energy, a more concentrated energy source, will become a reality within the next thirty years. In either case the starship will need the energy source to drive the accelerator rings and resulting thrust for periods up to perhaps a few months, then turn off for perhaps many years, and then resume operations for further maneuvers.

In the extreme case of travel to another galaxy, the energy source will need to turn off for up to millions of years of starship time. While the energy source is turned off, a residual "battery" will need to operate to support monitoring the progress of time, activating the startup of the main energy source, and possibly to detect and monitor objects ahead in the line of flight. This battery source may well be a plutonium source similar to those used in current space probes.[50]

The main energy source, if it is a nuclear reactor of some kind, will probably have to be a reactor that is different from current nuclear reactors. The nuclear fuel will have to be suspended in a liquid that becomes concentrated when the reactor is operating and becomes dilute when the reactor is brought to "stop." The dilution factor will determine the rate of nuclear reactions in the reactor. Since the startup process from a battery driven state needs to be gradual due to a "small" battery, it appears the nuclear reactor would be composed of perhaps five reactors of increasing size. The battery starts the smallest reactor by

---

[50] We note that a natural nuclear reactor existed in Central Africa for millions of years. (Parenthetical note: could this be the stimulus for the rapid evolution of species in Africa including early Mankind.)

concentrating its nuclear fuel. The smallest reactor then generates the energy to concentrate that fuel, and start the second smallest reactor, and so on until the main reactor is started. At this point the accelerators power up and the starship thrust begins to occur.

If the source of the energy is fusion energy then the startup process might begin in small stages in the boot up of the fusion reaction through fusing larger and larger amounts of (perhaps) He^3 with increasingly powerful laser beams a la the tokamuk approach.

During the coasting period of a starship, the nuclear reactors should be powered down to conserve nuclear fuel. When powering down a nuclear reactor based power source the nuclear material (U^{235} or plutonium) (the reactor fuel) residing in the liquid medium of the largest reactor would be diluted to sharply reduce fission reactions to "near zero" using the energy of the next largest reactor. Then this reactor would be similarly powered down by dilution using energy from the third largest reactor and so on until all the nuclear reactors are powered down. The smallest reactor would be powered down by a battery. This battery would retain enough energy to bring the smallest reactor back up after the coasting period ends. Then the reactors would boot up in turn to provide energy to the vehicle. (The battery would be at extremely low temperature during a coasting phase and thus not lose a significant amount of electrical power.)

In the case of a fusion power source a battery could be used to initiate the fusion power. A gradual turnoff process could execute at the start of a coasting phase to bring the fusion process to zero in such a way that the battery could initiate the boot up process for the fusion power source at the end of a coasting period.

REFERENCES

Blaha, S., 2004, *Quantum Big Bang Cosmology: Complex Space-time General Relativity, Quantum Coordinates, Dodecahedral Universe, Inflation, and New Spin 0, ½, 1 & 2 Tachyons & Imagyons* (Pingree-Hill Publishing, Auburn, NH, 2004).

Blaha, S., 2006, *A Unified Quantitative Theory Of Civilizations and Societies: 9600 BC - 2100 AD* (Pingree-Hill Publishing, Auburn, NH, 2006)

Blaha, S., 2007a, *Physics Beyond the Light Barrier: The Source of Parity Violation, Tachyons, and A Derivation of Standard Model Features* (Pingree-Hill Publishing, Auburn, NH, 2007).

Blaha, S., 2007b, *The Origin of the Standard Model: The Genesis of Four Quark and Lepton Species, Parity Violation, the ElectroWeak Sector, Color SU(3), Three Visible Generations of Fermions, and One Generation of Dark Matter with Dark Energy* (Pingree-Hill Publishing, Auburn, NH, 2007).

Blaha, S., 2008, *A Complete Derivation of the Form of the Standard Model With a New Method to Generate Particle Masses SECOND EDITION* (Pingree-Hill Publishing, Auburn, NH, 2008)

Blaha, S., 2009, *Bright Stars, Bright Universe* (Pingree-Hill Publishing, Auburn, NH, 2009)

Freeman, Marsha, *Krafft Ehricke's Extraterrestrial Imperative* (Apogee Books, 2009).

Huang, K., 1992, *Quarks, Leptons & Gauge Fields Second Edition* (World Scientific, River Edge, NJ, 1992).

Huang, K., 1998, *Quantum Field Theory* (John Wiley, New York, 1998).

Mallove, E. F. and Matloff, G. L., (1989) *The Starflight Handbook* ((John Wiley, New York, 1989).

Schmidt, S. and Zubrin, R. (eds.), 1996, *Islands in the Sky* (John Wiley, New York, 1996).

Weinberg, S., 1972, *Gravitation and Cosmology* (Wiley, New York, 1972).

Zubrin, R., 2000, *Entering Space* (Penguin Putnam, New York, 2000).

About the Author

Stephen Blaha is an internationally known physicist with extensive interests in Science, the Arts, and Technology. He received his Ph.D. in Theoretical Physics from The Rockefeller University (NY). He has written a highly regarded book on physics, consciousness and philosophy – *Cosmos and Consciousness*, a book on Science and Religion entitled *The Reluctant Prophets*, a book applying physics concepts to the history of civilizations, and books on Java and C++ programming. He developed a mathematical theory of civilizations that is described in *The Life Cycle of Civilizations*. Recently he completed a major new study of Cosmology: *Quantum Big Bang Cosmology: Complex Space-time General Relativity, Quantum Coordinates, Dodecahedral Universe, Inflation, and New Spin 0, ½, 1 & 2 Tachyons & Imagyons*. He has served on the faculties of several major universities. He was an Associate of the Harvard Physics Faculty for twenty years (1983-2003). He was also a Member of the Technical Staff at Bell Laboratories, a member of management at the Boston Globe Newspaper, a Director at Wang Laboratories, President of Blaha Software Inc and Janus Associates Inc. (NH), and 2008 Program Chair of the International Society for the Comparative Study of Civilizations.

Among other achievements he was a co-discoverer of the "r potential" for heavy quark binding developing the first (and still the only demonstrable) non-abelian gauge theory with an "r" potential; first suggested the existence of topological structures in superfluid He-3; first proposed Yang-Mills theories would appear in condensed matter phenomena with non-scalar order parameters; first developed a grammar-based formalism for quantum computers and applied it to elementary particle theories; first developed a new form of quantum field theory without divergences (thus solving a major 60 year old problem that enabled a unified theory of the Standard Model and Quantum Gravity without divergences to be developed); first developed a formulation of complex General Relativity based on analytic continuation from real space-time; first developed a generalized non-homogeneous Robertson-Walker metric that enabled a quantum theory of the Big Bang to be developed without singularities at t = 0; first generalized Cauchy's theorem and Gauss' theorem to complex curved multi-dimensional spaces; first developed a physically acceptable theory of faster-than-light particles – tachyons – of any spin; first showed a universe with three complex spatial dimensions has an icosahedral symmetry; first developed the form of the composition of extrema in the Calculus of Variations; first suggested that inflationary periods in the history of the universe were not needed; first proved Gödel's Theorem implies Nature must be quantum, first derived the form of the Standard Model from a complex extension of Special Relativity, first developed a quantitative harmonic oscillator-like model of the life cycle, and interactions, of civilizations, and first developed a new quantum-like form of Logic that eliminates Logical paradoxes including Gödel's Undecidability Theorem and generalizes to provide a fundamental basis for the Standard Model of Elementary Particles.

Blaha was also a pioneer in the development of UNIX for higher speed, financial and scientific applications, database benchmarking, and networking (1982); in the development of Desktop Publishing (1980's); and developed a hybrid shell programming technique (1982) that was a precursor to the PERL programming language. He received Honorable Mention in the Gravity Research Foundation Essay Competition in 1978, and was nominated for three "Awards for Technical Excellence" in 1987 by PC Magazine for PC software products that he designed and developed.

www.ingramcontent.com/pod-product-compliance
Lightning Source LLC
Chambersburg PA
CBHW082009190326
41458CB00010B/3126